Checklist of
Library
Building
Design
Considerations

ALA Editions purchases fund advocacy, awareness, and accreditation programs for library professionals worldwide.

Checklist of
Library
Building
Design
Considerations

6TH EDITION

WILLIAM W. SANNWALD

**An imprint of the
American Library Association**
Chicago   2016

William W. Sannwald was assistant to the city manager and manager of library design and development from 1997 to 2004, and was city librarian of the San Diego Public Library from 1979 to 1997. He is now a full-time faculty member in the business school at San Diego State University, where he teaches senior and MBA classes in management. He also works as a library building and administrative consultant. He is the author of numerous books and articles on library architecture and management and has presented papers at national and international conferences. Past president of the Library Administration and Management Association (LAMA), Sannwald was twice a jury member of the joint American Library Association/American Institute of Architects awards. He is the recipient of the San Diego American Institute of Architects chapter's highest honor, the Irving Gill Award, for his contributions to library architecture.

ISBN: 978-0-8389-1371-0 (paper)

Library of Congress Cataloging-in-Publication Data

Names: Sannwald, William W., author.
Title: Checklist of library building design considerations / William W. Sannwald.
Description: Sixth edition. | Chicago : ALA Editions, an imprint of the American Library
    Association, 2016. | Includes bibliographical references and index.
Identifiers: LCCN 2015028715 | ISBN 9780838913710 (print : alk. paper)
Subjects: LCSH: Library architecture--United States. | Library buildings—United States—
    Design and construction.
Classification: LCC Z679.2.U54 S36 2016 | DDC 727/.80973—dc23 LC record available at
    http://lccn.loc.gov/2015028715

Cover design by Kimberly Thornton. Building illustration © Vector House/Shutterstock, Inc.
Text design and composition by Kirstin McDougall in the Fira Sans and New Century
Schoolbook LT Std typefaces.

♾ This paper meets the requirements of ANSI/NISO Z39.48–1992 (Permanence of Paper).

Printed in the United States of America
20 19 18 17 16   5 4 3 2 1

Contents

7.   Compliance with ADA Accessibility Guidelines

8. Telecommunications, Electrical, and Miscellaneous Equipment

9. Interior Design and Finishes

Preface

This sixth edition of the *Checklist of Library Building Design Considerations* is published to accomplish a number of goals:

- ☑ To assist librarians, architects, administrators, and other members of a building design team in programming library spaces.

- ☑ To serve as a guide during the various stages of the design process in order to make sure that all needed spaces and functions are included in the library design.

- ☑ To enable the evaluation of existing library spaces as part of a library's Needs Assessment Process.

- ☑ To provide data and support to the library in its presentations to governing authorities and stakeholder groups.

In the *Checklist*, questions are asked concerning almost every aspect of space and function in a library building. The purpose of the questions is to make sure that the building design team in the evaluation and programming of spaces overlooks no element of the building. While the list of questions is probably not exhaustive, answering the questions in this document should ensure that no major design elements have been overlooked.

The *Checklist* is a valuable tool for programming and planning existing and potential library buildings. Most of the basic areas listed in the Checklist apply to college and university, community college, public, school, and special libraries. It should be relatively easy to adapt the Checklist to meet the requirements of almost any type of library.

The first edition was adapted from a manuscript produced by doctoral students in the School of Library and Information Studies at Texas Woman's University in Denton and was a collaborative product with the Library Building and Equipment sections of the Architecture for Public Libraries Committee.

All sections in the sixth edition have been reviewed, revised, and updated. In addition, some new chapters and sections are included:

☑ A new chapter on construction alternatives. Libraries often need to evaluate a brand-new building against an addition to an existing building, or a rehabilitation and/or renovation of an existing building. The chapter also examines preservation issues that may arise in older buildings and is an important issue in historically significant buildings. Other alternatives to new construction are considered, including virtual libraries and modular facilities.

☑ A new section dealing with community planning issues is included in the "Library Site Selection" chapter. These issues need to be considered in planning a library project in order to comply with campus and community rules and regulations.

☑ A new section in the "Interior Design and Finishes" chapter dealing with "Plus-Friendly Spaces." This section was inspired by Lori Smith's article in *American Libraries* about serving plus-sized patrons and staff. The plus-sized community presents a number of issues that library planners have to consider from labeling the weight that a stool is able to support to why wall hanging toilets (that allow easy floor cleaning) may be inappropriate for all rest rooms because a person's weight might rip the toilets from the wall.

☑ A brand new chapter dealing with "Entrepreneurial and Collaborative Spaces." This chapter represents some of the changes that have accelerated in libraries since the last edition. Included are makerspaces that seem to be thriving in academic, public, and school libraries as part of the "do it yourself" (DIY) movement. Another trend is "virtual business spaces" that are transforming the library into an office and meeting site for entrepreneurs and small businesspeople. The growth of lending non-traditional items has increased, including technology (video projectors, iPads, etc.), musical instruments, tools, and seeds. All of these activities change the traditional space allocations in libraries as well as the look and architectural atmosphere of the buildings.

☑ A new section under the "Building Systems" chapter that deals with the "Internet of Things." Everything in the library has the capability to be connected, including security cameras, electronic peripherals such as projectors, sensors (RFID and badge readers), wearable devices (smart watches, headsets), controllers for lighting and HVAC systems, and physical security (locks and gates).

☑ In the "Maintenance of Library Buildings and Property" chapter are two new sections. One provides information about building preventive maintenance and the other deals with building cleaning.

☑ In the "Building Occupancy and Post-Occupancy Evaluation" chapter is a new section dealing with certificates of occupancy. Many libraries have been kept from opening because they failed to get a certificate of occupancy from the local building officials. This is usually the result of the library's planning team (staff, architects, contractors) not interacting with local building officials throughout the project and the library being surprised at the end of construction.

In 2004 I took early retirement from the city of San Diego where I had served for twenty-four years as library director and assistant to the city manager. I was asked to teach full-time in the Management Department of the business school at San Diego State University, where I had taught part-time while I worked at the San Diego Public Library. I teach senior and MBA students in a variety of courses including organizational behavior, strategic management, performance management, corporate governance, and business ethics. Teaching is a great experience, and helps me as I continue to be involved in libraries through consulting, speaking, and writing.

I would like to dedicate this sixth edition to four librarians who most influenced me as a person and librarian:

- ☑ Irma Schlemmer, was the head librarian of the Mount Prospect (Illinois) Public Library who suggested librarianship as a career after I graduated from college and aided me in getting an H.W. Wilson Scholarship.

- ☑ Robert Rohlf, the best library-building consultant that I have worked with during my career.

- ☑ Clara S. Jones, who developed me as a library administrator when I worked for her at the Detroit Public Library during her two years as president of the American Library Association.

- ☑ Anna M. Tatar, a friend and dedicated colleague at the San Diego Public Library who made many changes happen at that library. Anna understands the power that libraries have to change society.

This publication should be viewed as a living document, and all comments and additions suggested for future editions are welcome. Please send them to:

William W. Sannwald
3538 Paseo Salamoner
La Mesa, CA 91941
Sannwald@gmail.com

1

Building Planning and Architecture

A. Indicators of Dissatisfaction with Existing Facilities

1. Has the mission of the library changed?

   □ □ □

   Comments: _____

   _____

2. Has the population served by the library increased or decreased?

   □ □ □

   Comments: _____

   _____

3. Have the demographics of the population served by the library changed?

   □ □ □

   Comments: _____

   _____

4. Has the library formed a partnership or alliance with another institution that requires a change in the physical building?

   □ □ □

   Comments: _____

   _____

5. Are there problems with the physical condition of the building (outdated systems, inflexible floor plans, ADA problems, difficulty in installing technology)?

   □ □ □

   Comments: _____

   _____

6. Does the existing building hinder the delivery of good service?

   □ □ □

   Comments: _____

   _____

	YES	NO	N/A

7. Is there enough space for the products and services the library offers? ☐ ☐ ☐

   Comments: _____

   _____

8. Has the mix of the products and services offered by the library changed? ☐ ☐ ☐

   Comments: _____

   _____

9. Does the physical building have the required infrastructure to accommodate current and future technology? ☐ ☐ ☐

   Comments: _____

   _____

10. In order to accommodate collection growth, have seats been exchanged for stacks? ☐ ☐ ☐

    Comments: _____

    _____

11. Is the atmosphere of the library pleasing for customers and staff? ☐ ☐ ☐

    Comments: _____

    _____

B. Institutional Planning Team

1. Has an institutional library planning team been formed? ☐ ☐ ☐

   Comments: _____

   _____

2. Who are the members of the library planning team:
   a) A representative of the legal owner (university, school, city, etc.)? ☐ ☐ ☐
   b) Library representatives? ☐ ☐ ☐
   c) Users (faculty, students, citizens, etc.)? ☐ ☐ ☐
   d) Other representatives with technical skills such as engineering, legal, financial, architectural, buildings, and so on? ☐ ☐ ☐
   e) Others (Friends of the Library, library committee members, faculty, etc.)? ☐ ☐ ☐

   Comments: _____

   _____

3. Will the architect hold a charrette for all members of the library planning team?
   A charrette is essentially a design workshop where designers, residents, developers, city officials, university representatives, planners, and other interested parties come together to envision what the planners hope that a new library will accomplish.
   It is a short-term, intense design tool to flesh out the planning team's vision for the future. ☐ ☐ ☐

   Comments: _____

   _____

4. What roles will members of the library planning play:

   a) Advising (gathering and disseminating information about the project)? ☐ ☐ ☐

   b) Innovating (suggesting new ideas or new ways of tackling old problems)? ☐ ☐ ☐

   c) Promoting ("selling" the project to interested stakeholders)? ☐ ☐ ☐

   d) Developing (assessing and developing ideas for practical implementation)? ☐ ☐ ☐

   e) Maintaining (ensuring that the infrastructure is in place so that the team can work with maximum efficiency)? ☐ ☐ ☐

   f) Linking (coordinating all work roles to ensure maximum cooperation and interchange of ideas, expertise, and experience)? ☐ ☐ ☐

   Comments: _____

   _____

5. Who will be the spokesperson and chief contact for the institution on the project? It is important that only one person speak for the institution during all the stages of the building process. ☐ ☐ ☐

   Comments: _____

   _____

6. How will conflict be resolved on the project? ☐ ☐ ☐

   Comments: _____

   _____

7. Who will make the final decision on design, space allocations, costs, and change orders? ☐ ☐ ☐

   Comments: _____

   _____

C. Determining Space Needs

1. Has the library staff and administration met to decide the mission and long-term vision of the library? Space allocations and needs should be based on the vision and mission of the library. ☐ ☐ ☐

   Comments: _____

   _____

2. What is the useful life of the new building? Most building changes should accommodate the library for a period of approximately twenty years. ☐ ☐ ☐

   Comments: _____

   _____

3. If a building change is planned that is an interim solution, how will this impact future needs? ☐ ☐ ☐

   Comments: _____

   _____

4. What existing programs will be discontinued in the new building? ☐ ☐ ☐

   Comments: _____

   _____

5. What new programs will be added in the new building? ☐ ☐ ☐

   Comments: _____

   _____

6. What will be the growth of staff over the next twenty years? ☐ ☐ ☐

   Comments: _____

   _____

7. What changes will take place in the service population over the next twenty years? ☐ ☐ ☐

   Comments: _____

   _____

8. What will be the growth of the collection over the next twenty years? ☐ ☐ ☐

   Comments: _____

   _____

9. What technology will be required to support library programs over the next twenty years? ☐ ☐ ☐

   Comments: _____

   _____

10. In estimating the size of the new building, have all of the following four factors been considered?

    a) "Building up" library spaces based on the programs and activities that the library wants to undertake in the new building. The total of all spaces equals the ideal size of the new building or expansion. ☐ ☐ ☐

    b) What is the size of library buildings in similar institutions? It is valuable to have a database of ten similar libraries that may be consulted not only for facility size, but also for other measurable aspects of library space. If you can't measure it, you can't manage it. ☐ ☐ ☐

    c) What do library association, regional, state, and other guidelines and standards call for as far as space guidelines? ☐ ☐ ☐

    d) What can the library afford? This sometimes is the deciding factor. ☐ ☐ ☐

    Comments: _____

    _____

11. Has a library building consultant been hired to help the library in planning? ☐ ☐ ☐

    Comments: _____

    _____

12. Has a building program been prepared detailing space needs, adjacencies, and unique functions and features of the proposed building? ☐ ☐ ☐

    Comments: _____

    _____

13. Has the library building consultant prepared the program or advised staff on preparing the program? ☐ ☐ ☐

    Comments: _____

    _____

	YES	NO	N/A

14. What will be the growth of seating requirements over the next twenty years? ☐ ☐ ☐

   Comments: _____

   _____

15. Have the Association of College and Research Libraries' (ACRL's) "Standards for Libraries in Higher Education" been consulted? (www.ala.org/acrl/standards/ standardslibraries) ☐ ☐ ☐

   Comments: _____

   _____

16. While ACRL does not have any specific national quantitative standards, some questions for designers of library space are available:

   a) Does the library provide well-planned, secure, and sufficient space to meet the perceived needs of staff and users? ☐ ☐ ☐

   b) Are building mechanical systems properly designed and maintained to control temperature and humidity at recommended levels? ☐ ☐ ☐

   c) What are the perceptions of users regarding the provision of conducive study spaces, including a sufficient number of seats and varied types of seating? ☐ ☐ ☐

   d) Is there enough space for current library collections and future growth of print resources? ☐ ☐ ☐

   e) Does the staff have sufficient workspace, and is it configured to promote efficient operations for current and future needs? ☐ ☐ ☐

   f) Does the library's signage facilitate use and navigation of the facilities? ☐ ☐ ☐

   g) Does the library provide ergonomic workstations for its users and staff? ☐ ☐ ☐

   h) Are electrical and network wiring sufficient to meet the needs associated with electronic access? ☐ ☐ ☐

   i) Does the library meet the requirements of the Americans with Disabilities Act (ADA)? ☐ ☐ ☐

   j) Are facilities provided to distance learners considered in the context of the ACRL's *Guidelines for Distance Learning and Library Services* (www.ala.org/ acrl/guides/distlrng.html)? ☐ ☐ ☐

   Comments: _____

   _____

17. Has the ACRL/LAMA "Guide for Architects in Planning Higher Education Library Spaces" been consulted (http://wikis.ala.org/acrl/index.php/ACRL/LAMA_Guide _for_Architects)? ☐ ☐ ☐

   Comments: _____

   _____

18. Have Public Library Association planning documents been consulted?

   a) *Public Library Service Responses, 2007.* These 18 service responses are an update of the original 13 service responses that were published in the 1997 publication, *Planning for Results: A Public Library Transformation Process.* Each service response contains eight sections: the title, the description, suggested target audiences, typical services and program in libraries that select this as a priority, potential partners, policy implications, critical resources, and possible measures. Each of the service responses suggests physical needs required to provide the response. ☐ ☐ ☐

b) Has the 1962 "Interim Standards for Small Public Libraries" been consulted? (This standard has never been rescinded and is the only standard that recommends quantitative measures for public library size; 0.07 sq. ft./capita). ☐ ☐ ☐

c) Has *Managing Facilities for Results: Optimizing Space for Services* by Cheryl Bryan been consulted? This book dovetails with the basics outlined in *The New Planning for Results: A Streamlined Approach* to help public libraries plan physical spaces. ☐ ☐ ☐

Comments: _____

_____

19. Have the following sources for school library media centers been consulted?

a) The American Association of School Librarians has many excellent facilities planning suggestions on its website: www.ala.org/aasl/standards-guidelines ☐ ☐ ☐

b) Margaret Sullivan, *Library Spaces for 21st-Century Learners: A Planning Guide for Creating New School Library Concepts.* Focuses on planning contemporary school library spaces with user-based design strategies. The book walks school librarians and administrators through the process of gathering information from students and other stakeholders involved in planning a resource-rich learning space. ☐ ☐ ☐

c) Rolf Erikson and Carolyn Markuson, *Designing a School Library Media Center for the Future.* This book will help school librarians anticipate needs and participate in the planning process with the architect, consultant(s), building committee, and administrators. ☐ ☐ ☐

d) Has Thomas L. Hart's *The School Library Media Facilities Planner* been consulted? This book is full of suggestions for all stages of the facility planning process, whether new construction or remodeling. The book offers examples of exemplary facilities, success stories, and problems encountered. It includes appendixes for a glossary, model policies and planning documents, and a companion DVD that takes the reader on tours of several new or remodeled school libraries. ☐ ☐ ☐

Comments: _____

_____

20. Has the Library Leadership and Management Association's (LLAMA's) *Building Blocks for Planning Functional Library Space* (3rd ed.) been consulted? A great source for specifying spaces required for library furniture and equipment. ☐ ☐ ☐

Comments: _____

_____

D. Joint Use Considerations

1. Is there another organization and/or department on the campus or in the community that may offer synergy to the library by sharing facilities? ☐ ☐ ☐

Comments: _____

_____

	YES	NO	N/A

2. Is there another library or other organization that may offer potential synergy for a joint use facility? ☐ ☐ ☐

   Comments: _____

   _____

3. Do the missions of the libraries considering a joint facility have enough commonalities to enhance the chances of success? ☐ ☐ ☐

   Comments: _____

   _____

4. Are there possible efficiency and cost savings by having a joint facility? ☐ ☐ ☐

   Comments: _____

   _____

5. Can the quality and quantity of service be improved for both libraries through a joint facility? ☐ ☐ ☐

   Comments: _____

   _____

6. If a joint facility is agreed to, has a joint interagency agreement been negotiated? Some of the factors to be considered in the agreement are:

   a) *Governance.* A written agreement is strongly recommended, and the agreement should list the parties entering into the agreement. The agreement should provide a clear demarcation of responsibility. ☐ ☐ ☐

   b) *Funding.* It is important to determine and put into an agreement the financial responsibilities of each party. This includes both capital and operational costs. ☐ ☐ ☐

   c) *Ownership of assets.* The agreement should clarify the ownership of assets brought into the shared library (such as equipment, collection, etc.) and how ownership will be determined in the event of termination of the combined library agreement. ☐ ☐ ☐

   d) *Hours of operation.* The agreement should list the hours of operation of both libraries, and if either partner has restrictions on use. ☐ ☐ ☐

   e) *Staffing.* Because a combined library is two libraries sharing one facility, it is recommended that local staffing requirements for both types of libraries should be met. The two staffs may have different certification or licensing requirements. ☐ ☐ ☐

   f) *Volunteers.* Some libraries rely on youth and parent volunteers, and the other library may not use as many volunteers. ☐ ☐ ☐

   g) *Collections.* Care must be taken to develop collections that are responsive to the needs of both sets of users. ☐ ☐ ☐

   h) *Changes.* How will changes in any of the above policies be determined? The agreement must be flexible enough to allow modifications as conditions change. ☐ ☐ ☐

   i) *Termination of the agreement.* If for some reason a termination is desired in the future, the agreement should state the conditions related to termination of the agreement. ☐ ☐ ☐

   Comments: _____

   _____

E. Selecting a Library Building Consultant

1. Is there someone on the staff who has the necessary planning knowledge and experience of the functional needs and requirements of library buildings? If not, a library building consultant should be retained.

   □ □ □

   Comments: _____
   _____

2. Has the consultant been retained at the very start of the building planning process so that he or she can take part in every step of the project?

   □ □ □

   Comments: _____
   _____

3. Is the consultant listed in the Library Consultants Directory? (www.libraryconsultants.org/submail.html)?

   □ □ □

   Comments: _____
   _____

4. Does the consultant have broad and diversified technical experience in planning new buildings, renovations, additions, and conversion of other buildings into library buildings?

   □ □ □

   Comments: _____
   _____

5. Does the consultant have the organizational and record-keeping skills needed to document and respond to key events and activities during the project?

   □ □ □

   Comments: _____
   _____

6. Does the consultant have the personal characteristics, experience, and skills necessary to assist a library in its unique planning and building needs?

   □ □ □

   Comments: _____
   _____

7. Does the consultant have the written and verbal communication skills required to interact with all stakeholders?

   □ □ □

   Comments: _____
   _____

8. Does the consultant have the political skills necessary to listen and respond to the concerns of all who may have a stake in the building project?

   □ □ □

   Comments: _____
   _____

9. Does the building consultant have the ability to explain a point of view and to persuade others of the importance of carrying out the consultant's recommendations?

   □ □ □

   Comments: _____
   _____

10. Will the consultant provide advice on the selection of the architect and other members of the building's technical planning team? ☐ ☐ ☐

    Comments: _____

    _____

11. Is the consultant's schedule flexible enough for him or her to be available for meetings with the library's planning committee when required? ☐ ☐ ☐

    Comments: _____

    _____

12. Is the consultant available by telephone, surface mail, or electronic communication to answer questions and provide guidance when his or her physical presence is not required? ☐ ☐ ☐

    Comments: _____

    _____

F. Choosing an Architect

1. Does the library staff play a major role in selection of the architect? ☐ ☐ ☐

   Comments: _____

   _____

2. Has the group responsible for selection of the architect developed selection criteria? ☐ ☐ ☐

   Comments: _____

   _____

3. Does the architectural selection process include:

   a) Announcement of the proposed project in an official publication used by the client organization or in the general press? ☐ ☐ ☐
   b) Requests for proposals and/or information? ☐ ☐ ☐
   c) Submittals by interested firms? ☐ ☐ ☐
   d) Provision of standardized forms so that a uniform evaluation of firms may be used during the evaluation process? ☐ ☐ ☐
   e) Evaluation based on the selection criteria developed by the group responsible for selection of the architect? ☐ ☐ ☐
   f) Interviews with the "short list" of firms that the selection group has decided best meets the selection criteria? ☐ ☐ ☐
   g) A tour of the site or facility organized prior to the final selection of the architectural design team? It may be appropriate for the tour to be arranged prior to the short-listing process, or it may be considered more appropriate to delay the tour until after a final list of architect finalists is determined. ☐ ☐ ☐
   h) Ranking of the top firms to identify the best-qualified firms? ☐ ☐ ☐
   i) Selection of the top-ranked firm based on the interview discussions and the selection criteria? ☐ ☐ ☐
   j) Notification of unsuccessful firms, and a debriefing as to why they were not selected? ☐ ☐ ☐

   Comments: _____

   _____

4. While not necessarily recommended, does the selection process involve:

   a) Limited or open architectural competitions? ☐ ☐ ☐
   b) Design/build competitions? ☐ ☐ ☐
   c) Bidding among various competitors? ☐ ☐ ☐

   Comments: _____

   _____

5. Is the architectural firm an individual, partnership, corporation, or joint venture? ☐ ☐ ☐

   Comments: _____

   _____

6. Will the person who presents for the architectural team be involved in the project? ☐ ☐ ☐

   Comments: _____

   _____

7. Who is the person who will be in charge of designing the project? ☐ ☐ ☐

   Comments: _____

   _____

8. Who is the person who will supervise the project from design to completion? ☐ ☐ ☐

   Comments: _____

   _____

9. Is the architect or architectural firm registered to practice in the state? ☐ ☐ ☐

   Comments: _____

   _____

10. Is the architect of record registered to practice in the state? ☐ ☐ ☐

   Comments: _____

   _____

11. Are all key personnel and sub-consultants involved in the project from the architect's office identified? ☐ ☐ ☐

   Comments: _____

   _____

12. Are the architect's support team members identified: the landscape architect, civil engineer, structural engineer, acoustic engineer, mechanical engineer, electrical engineer, ADA compliance officer, and any other key specialists involved in the project? ☐ ☐ ☐

   Comments: _____

   _____

13. Does the architectural team have the required support equipment—computer-aided design (CAD), 3D modeling, renderings in house, and so on? ☐ ☐ ☐

   Comments: _____

   _____

	YES	NO	N/A

14. Are all members of the architect's support team part of the firm, or does the architect retain them as sub-consultants? ☐ ☐ ☐

Comments: _____

_____

15. Do the architect's workload and organization provide enough resources to devote time and energy to the project? ☐ ☐ ☐

Comments: _____

_____

16. Does the architect have experience in working with public agencies? ☐ ☐ ☐

Comments: _____

_____

17. Does the architect have prior experience in designing libraries? In some cases, it may be advantageous to have an architect who has not worked on a library building. ☐ ☐ ☐

Comments: _____

_____

18. If the architect has not worked with libraries, does the architect have a plan to become knowledgeable about library needs? This may require a library building consultant preferably retained by the client. ☐ ☐ ☐

Comments: _____

_____

19. Is the architect an empathetic listener, willing to understand library needs? ☐ ☐ ☐

Comments: _____

_____

20. How will the architect gather information about library operations, project site, and so forth? ☐ ☐ ☐

Comments: _____

_____

21. What is the architect's design philosophy? ☐ ☐ ☐

Comments: _____

_____

22. Will the architect place library needs before design considerations? ☐ ☐ ☐

Comments: _____

_____

23. Does the architect's workload allow the firm to devote adequate time to the project? ☐ ☐ ☐

Comments: _____

_____

	YES	NO	N/A

24. Does the architect have solid reference reports from past clients? ☐ ☐ ☐

    Comments: _____

    _____

25. In projects completed by the architect:

    a) Did the projects come in at or under budget? ☐ ☐ ☐
    b) Did the projects come in on time? ☐ ☐ ☐
    c) What is the extent of change orders in number and dollars? ☐ ☐ ☐
    d) If there have been change orders, has it been determined whose fault they
       were? (Not all change orders are the architect's fault.) ☐ ☐ ☐
    e) What litigation has occurred against the architect? ☐ ☐ ☐
    f) What litigation has occurred against the architect's former clients by
       the architect? ☐ ☐ ☐

    Comments: _____

    _____

26. Does the architect have written and verbal communication skills required for
    interacting with all stakeholders? ☐ ☐ ☐

    Comments: _____

    _____

27. Does the architect have the political skills necessary to listen and respond to the
    concerns of all external and internal building-project stakeholders? ☐ ☐ ☐

    Comments: _____

    _____

28. Does the architect have the ability to explain the reasons for a point of view and to
    persuade others of the importance of carrying out his or her recommendations? ☐ ☐ ☐

    Comments: _____

    _____

29. Does the architect's proposed fee fit into the fee guidelines of the American Institute
    of Architects? ☐ ☐ ☐

    Comments: _____

    _____

30. Is the architect's proposed fee within the library's budget? ☐ ☐ ☐

    Comments: _____

    _____

G. Choosing a Contractor

1. Will the award of the construction contract be made by a competitive bidding process? ☐ ☐ ☐

   Comments: _____

   _____

YES | NO | N/A

2. Is a call or invitation to bid advertised in an official publication used by the client organization or in the general press? ☐ ☐ ☐

   Comments: _____

   _____

3. For purposes of soliciting bids and awarding a contract, has the library declared who the "owner" is? (Usually the owner has legal and financial jurisdiction over the operations of the library.) ☐ ☐ ☐

   Comments: _____

   _____

4. Does the bidding period extend for a period of four to six weeks so that potential bidders may prepare their bids? ☐ ☐ ☐

   Comments: _____

   _____

5. Are standardized bid forms provided so that a uniform evaluation of contractors may be used during the bid evaluation process? ☐ ☐ ☐

   Comments: _____

   _____

6. Are the architect and a library representative available to answer technical questions from potential bidders during the bid period? ☐ ☐ ☐

   Comments: _____

   _____

7. Has a time and place been specified for opening bids? ☐ ☐ ☐

   Comments: _____

   _____

8. During the bid opening, are all bids made public? ☐ ☐ ☐

   Comments: _____

   _____

9. After bids are received, are they "taken under advisement" by the owner so that the bids may be analyzed? ☐ ☐ ☐

   Comments: _____

   _____

10. During the bid analysis period, and before the contract is awarded, is the lowest bidder checked for responsibility and:

    a) Is the bid submitted complete, accurate, and in compliance with the requirements, drawings, and specifications provided by the owner? ☐ ☐ ☐
    b) Is the contractor a licensed and registered contractor for your state? ☐ ☐ ☐
    c) Does the contractor have sufficient staff to execute the scope of the project? ☐ ☐ ☐
    d) Has the contractor been in business long enough to establish a "track record"? ☐ ☐ ☐
    e) What references does the contractor provide? ☐ ☐ ☐

f) What is the contractor's record in successfully completing other projects? □ □ □

g) Does the contractor usually complete projects in the period specified? □ □ □

h) Has any litigation occurred against the contractor? □ □ □

i) What litigation has the contractor brought against previous clients and/or architects? □ □ □

j) What is the reputation of the subcontractors that the contractor has specified? □ □ □

k) Does the contactor have a change management process that is logical and easy to understand? □ □ □

l) Does the contractor have the necessary insurance and bonds to protect the owner as called for in the legal and financial specifications? □ □ □

m) Will the contractor obtain any necessary building or zoning permits? □ □ □

n) Does the contractor have the appropriate licenses to do the job? □ □ □

o) Does the contractor have a pleasant business manner and is the person comfortable to talk to? □ □ □

Comments: _____

_____

11. Is the bid awarded to the lowest responsible bidder? □ □ □

Comments: _____

_____

H. Architectural Design

1. Does the library design proposed by the architect meet the building program requirements? □ □ □

Comments: _____

_____

2. Does the design have the character and power to make the library building a focus for its community or campus? □ □ □

Comments: _____

_____

3. Does the design take full advantage of all positive features of the site? □ □ □

Comments: _____

_____

4. Does the design compensate to the best degree possible for the negative aspects of the site? □ □ □

Comments: _____

_____

5. Is the architectural character distinctive in appearance, yet in harmony with its surroundings? □ □ □

Comments: _____

_____

6. Does the design welcome users and encourage nonusers to enter and investigate the library?

   Comments: _____

   _____

7. Does the design create a building that is unmistakably public in character and function, yet very comfortable and non-intimidating for the user?

   Comments: _____

   _____

8. Is the interior design in harmony with the exterior of the library?

   Comments: _____

   _____

9. Do interior finishes create a space that is inviting to users, yet able to stand up to the wear and tear of heavy public use?

   Comments: _____

   _____

10. Does the design provide flexibility to take advantage of changes in library products and services, roles and activities, as well as technology?

    Comments: _____

    _____

11. Does the design consider light, books, people, and the surrounding space as integral to each other?

    Comments: _____

    _____

12. Does the design express symbolically the important values of knowledge and learning?

    Comments: _____

    _____

13. Does the design merchandise the products and services of the library by incorporating design features used successfully in retail merchandising?

    Comments: _____

    _____

14. Does the design solve the paradoxical needs within a library of spatial openness and seclusion by creating:

    a) The ability to orient oneself within the visible total enclosure yet feel anchored to a particular part of it?

    b) The possibility of easy supervision by staff without the sense of being left exposed in a large impersonal space?

    c) A gradation of different spaces within the library, ranging from open areas of public activity to alcoves of semiprivate activity?

d) Areas that have a sense of intimacy within the overall public setting? ☐ ☐ ☐

e) A wide variety of reading areas so that users have many choices to fit their mood or reading environment needs? ☐ ☐ ☐

f) A clear understanding upon entry to the library (and while moving within the library) of the general purpose of each library area? ☐ ☐ ☐

g) Clearly visible staff areas as a means for bringing information, services, and people together? ☐ ☐ ☐

Comments: _____

_____

15. Does the library design plan encourage efficient traffic patterns from outside the structure into the building? ☐ ☐ ☐

Comments: _____

_____

16. Does the library design plan encourage efficient traffic patterns within the building? ☐ ☐ ☐

Comments: _____

_____

17. Does the library design provide for the maximum use of self-service by the library's customers? ☐ ☐ ☐

Comments: _____

_____

18. Does the design reflect the unique natural climate of the region where it is located? ☐ ☐ ☐

Comments: _____

_____

19. Are windows treated or shaded to prevent the hot and damaging rays of the sun from penetrating the interiors? ☐ ☐ ☐

Comments: _____

_____

20. Does the design provide flexibility in the placement of lighting fixtures, air ducts and registers, electrical power, and communication linkages to provide long-term adaptability? ☐ ☐ ☐

Comments: _____

_____

21. Does the spacing of columns, shafts, and other architectural elements provide flexibility and the effective use of space? ☐ ☐ ☐

Comments: _____

_____

22. Does the modular system employed meet the unique space needs of the library? ☐ ☐ ☐

Comments: _____

_____

Building Construction Alternatives

A. New Construction Considerations

1. What is new library construction? To build library space that is designed to meet the stated needs of a library. Many of the following considerations may also apply to the other construction alternatives listed below. ☐ ☐ ☐

   Comments: _____

   _____

2. New construction considerations:

   a) Will a new library building mitigate the issues that the library has with the existing building? ☐ ☐ ☐
      i) Lack of space for collections, programs, and technology? ☐ ☐ ☐
      ii) Lack of space for library customers? ☐ ☐ ☐
      iii) Lack of space for library staff? ☐ ☐ ☐
      iv) Environmental hazards? ☐ ☐ ☐
      v) Physical and personal safety concerns? ☐ ☐ ☐
      vi) Lack of flexibility for making program or operational changes in the existing building? ☐ ☐ ☐
      vii) Uninviting atmospherics that do not attract library users? ☐ ☐ ☐
   b) Will the library have control or significant influence in the library's location? ☐ ☐ ☐
   c) Will the library control the design of the functional areas of the library? ☐ ☐ ☐
   d) Will the library be able to control selection of furniture, fixtures, and equipment? ☐ ☐ ☐
   e) Will the library be able to take advantage of newer technologies and materials with the new construction in order to provide better service and customer and staff satisfaction? ☐ ☐ ☐

f) Does the library site permit enough space for construction staging?
A construction staging area is a designated area where vehicles, supplies, and construction equipment are positioned for access and use to a construction site. It may be either on the library site or an adjacent area. ☐ ☐ ☐

g) Will the library be able to afford the cost of construction and furniture, fixtures and equipment? ☐ ☐ ☐

h) Will the library be able to afford to operate an expanded physical facility that may require more staffing and service points? ☐ ☐ ☐

i) Will a new building attract donors who may help finance construction and operating costs? ☐ ☐ ☐

j) Is the "pride of ownership" which occurs with a new library important to the school/community/university/institution? ☐ ☐ ☐

Comments: _____

_____

B. Building Addition Considerations

1. What is a building addition? To add new space to an existing building that may or may not match the style of the original. ☐ ☐ ☐

   Comments: _____

   _____

2. Building addition considerations:

   a) Did the original design of the building and site include a plan for expansion? ☐ ☐ ☐

   b) Is the existing building a good location for the library? If not, perhaps new construction or the acquisition of a building for rehabilitation should be considered. ☐ ☐ ☐

   c) Is the existing building structurally sound? ☐ ☐ ☐

   d) Is there sufficient space to accommodate an addition? ☐ ☐ ☐

   e) Is it possible for the addition to be a horizontal expansion? This will require additional land adjacent to the existing building. ☐ ☐ ☐

   f) Is it possible for the addition to be a vertical expansion? This will require the existing structure to accommodate the additional loads (weight) in adding vertical space. In most cases the expansion will consist of adding stories up, but some libraries have expanded under the existing building. ☐ ☐ ☐

   g) Will building codes allow an addition? ☐ ☐ ☐

   h) Will the addition also include renovation of existing spaces? ☐ ☐ ☐

   i) Will the new addition integrate library operations seamlessly with existing spaces? ☐ ☐ ☐

   j) Has there been consideration of not only how the addition will look from the outside, but also what the outside will look like as viewed from the addition? ☐ ☐ ☐

   k) Is there opposition to expanding the library due to historical or sentimental issues tied to the existing building? ☐ ☐ ☐

   Comments: _____

   _____

C. Rehabilitation/Renovation of Existing Buildings to Library Use Considerations

1. What is rehabilitation? The process of making possible a compatible use for a library through repair, alterations, and additions to a building designed for another function. In most cases, preserving those portions or features of the existing building which convey its historical, cultural, or architectural values is desirable. □ □ □

    Comments: _____

    _____

2. What are some libraries that have successfully located to rehabilitated spaces?

    a) The McAllen Public Library (Texas) main library, which opened in late 2011 in a former Walmart store, garnered many awards, including the coveted American Institute of Architects (AIA) Honor Award for Interior Architecture. □ □ □

    b) Library of Braunschweig University. The extension to the library building is constructed from parts of the Mexican pavilion from the EXPO 2000 and was designed by the well-known artist Ricardo Legorreta. □ □ □

    Comments: _____

    _____

3. Rehabilitation considerations

    a) Are there potential vacant buildings available for purchase at a comparatively low price? □ □ □

    b) Will reuse of the structure preserve an architecturally significant building for the campus or community? □ □ □

    c) Is the building a prominent and significant structure that will enhance the image of the library? □ □ □

    d) Will reuse of an existing building result in lower land, construction, and environmental costs than new construction? □ □ □

    e) Will building reuse preserve materials and craftsman-style construction at a fraction of the cost of new construction? □ □ □

    f) Is the vacant building in a location that is desirable for library needs? □ □ □

    g) Is the building structurally sound and safe? □ □ □

    h) Will the building be able to meet current building codes? □ □ □

    i) Will the building layout adapt to library programs and activities? □ □ □

    j) Will the building layout increase staff operating costs? □ □ □

    k) Will floors be able to support the weight of books and heavy library equipment? □ □ □

    l) Are there ADA accessibility issues that can't be easily remedied? □ □ □

    m) Will it be cost-effective to adapt building systems (HVAC, lighting, etc.) to the needs of the library? □ □ □

    n) Are there environmental conditions in the building that may not be easily remedied (asbestos, mold, rot, etc.)? □ □ □

    Comments: _____

    _____

D. Preserving Existing Library Buildings Considerations

1. What is preservation? Preservation is a treatment for keeping a building in its current state without any changes to the function or size of the building. ☐ ☐ ☐

    Comments: _____

    _____

2. Preservation considerations:

    a) Will it preserve a building that is of historic significance to the campus or community for future generations? ☐ ☐ ☐
    b) Will preservation solve the issues that have caused the library to consider making changes to its physical space? ☐ ☐ ☐
    c) Is it an environmentally responsible practice? By reusing existing buildings historic preservation is essentially a recycling program. ☐ ☐ ☐
    d) Is there an immediate advantage in preserving an older building? Yes in that a building already exists; therefore energy is not necessary to demolish a building or create new building materials. ☐ ☐ ☐
    e) Will aesthetics and craftsmanship in the building be possible to replicate today? ☐ ☐ ☐
    f) Will the refurbishment/replacement of existing historic features and fixtures exceed the library's budget? ☐ ☐ ☐
    g) Will preservation provide quicker occupancy than other building project alternatives? ☐ ☐ ☐
    h) Will upgraded building systems be difficult to install because of limitations of the building structure? ☐ ☐ ☐
    i) Will programs and services offered by the library be compatible with the renovated structure? ☐ ☐ ☐
    j) Will the expense of bringing a historic structure up to standards exceed the cost of new construction? ☐ ☐ ☐
    k) Will the historic building meet contemporary building codes? ☐ ☐ ☐
    l) Are there safety and accessibility issues that can't be remedied by preservation? ☐ ☐ ☐

    Comments: _____

    _____

E. Virtual Library Considerations

1. What is a virtual library? It is a digital space that keeps and organizes electronic books and their associated documents. It also can refer to a space where the books can be read. These spaces can include computers, mobile devices, and the Internet. The term *virtual library* also can be used to refer to a collection of books that are readily available to be read on the Internet. These books often can be read for free if they are available under the public domain. Many libraries have established virtual libraries to share their collections. ☐ ☐ ☐

    Comments: _____

    _____

2. Virtual library considerations:

a) Will the library have its entire collection in the virtual library? If all of a library's e-resources were cataloged to the virtual library, the collocation would be comprehensive. When users initiated a search limited to this "library" they would search across all of the library's electronic resources on all campuses or libraries. ☐ ☐ ☐

b) Will the library catalog only a portion of its collection in the virtual library? Do libraries that establish virtual libraries include all their materials in the collection? A library may decide to catalog only a portion of the e-resources to the virtual library and catalog others to individual libraries. Users would need to be aware of this distinction and appropriately direct or limit their searches. ☐ ☐ ☐

c) What authentication is required to access the collections? ☐ ☐ ☐

d) Will the library save physical space with a virtual library? An advantage of using a virtual library is that it is able to store many books in a small amount of space. Physical books and materials can take up a lot of space. By using a digital library, readers can access their entire personal libraries in one location, on either a computer or a mobile device. The need for physical books diminishes by using a virtual library; so digital libraries also have the benefit of helping to conserve paper resources. ☐ ☐ ☐

e) Will the library loan tablets and devices to access the electronic books? Space required to house electronic book readers might offset some of the saved space. ☐ ☐ ☐

Comments: _____

_____

F. Alternatives to New Library Construction

1. Has the library investigated outsourcing all or a portion of library services? ☐ ☐ ☐

Comments: _____

_____

2. Has the library investigated leased space for public and nonpublic sections and activities that could function effectively outside the library in another location? ☐ ☐ ☐

Comments: _____

_____

3. Has the library investigated adjacent buildings that might be acquired in order to add square footage to the existing library? ☐ ☐ ☐

Comments: _____

_____

4. Has the collection been weeded to eliminate unneeded books and media that take up space in the library? ☐ ☐ ☐

Comments: _____

_____

5. Has the library's programming been reviewed, and programs eliminated that are no longer required but which take up space in the library?

☐ ☐ ☐

Comments: _____

_____

6. Is it possible to install high-density stacks to provide more book storage within the exact same book stacks floor space area?

☐ ☐ ☐

Comments: _____

_____

7. Has the library considered remote book storage facilities? Remote storage/shelving facilities supplement the storage capacity of library facilities, and prolong the useful life of library materials in a preservation-sensitive environment away from the campus/community library. An example is the Buhr Shelving Facility operated by the University of Michigan that currently houses 2.5 million items. Libraries may also operate the facilities as an institution-only facility or a cooperative of like libraries. Users request delivery of stored materials and usually receive them on an agreed-upon schedule.

☐ ☐ ☐

Comments: _____

_____

8. Has the library considered a robotic book delivery system housed in the library? These systems can store millions of items in a climate-controlled environment and deliver any of them within minutes by a click in the online catalog. Most systems require one-ninth the space of conventional shelving, allowing more space in the library for users and services.

☐ ☐ ☐

Comments: _____

_____

9. Has the library investigated modular buildings and/or kiosks that might be used instead of new construction? Modular buildings are available from a number of sources, not just library vendors.

☐ ☐ ☐

Comments: _____

_____

10. Has a library e-branch been considered to provide supplemental services? This kiosk-type facility provides immediate library access wherever customers need it. Much like a bank's ATM, it may be installed at convenient locations throughout a community or campus. It provides a direct link to the library's database as well as Internet access 24 hours a day, 7 days a week. Everything from stock quotes to local weather to health care information can be immediately accessed at this freestanding library terminal.

☐ ☐ ☐

Comments: _____

_____

11. Has the library considered a freestanding lending library outside the library building? The lending library looks and functions as a vending machine and extends the reach of the library by providing popular materials in remote locations from your library including campus dorms, community centers, transportation centers, and so on. The lending library is integrated with the library system and is accessed with a library card. It may contain books and media. ☐ ☐ ☐

Comments: _____

_____

12. Has the library considered mobile units that take library services to the community or campus? ☐ ☐ ☐

Comments: _____

_____

13. Has the library considered supplementing its collections by using an eGranary Digital Library—also known as "the Internet in a box"? This service provides millions of digital educational resources to institutions lacking adequate Internet access. The digital library contains built-in tools for subscribers to upload and edit local materials as well as create and edit their own websites, which are stored locally. ☐ ☐ ☐

Comments: _____

_____

Library Site Selection

A. General Conditions

1. Is the site conveniently located to the population served by the library?   ☐ ☐ ☐

   Comments: _____

   _____

2. Does the site provide high visibility and identification to the population served?   ☐ ☐ ☐

   Comments: _____

   _____

3. Is the site affordable?   ☐ ☐ ☐

   Comments: _____

   _____

4. Will the site provide visibility of the building and its function from the street?   ☐ ☐ ☐

   Comments: _____

   _____

5. Will a library be an appropriate use of the land parcel in question?   ☐ ☐ ☐

   Comments: _____

   _____

6. Will the site retain or enhance the natural contours of the land?   ☐ ☐ ☐

   Comments: _____

   _____

7. Is the site zoned for a library? If not, is future library zoning possible? ☐ ☐ ☐

   Comments: _____

   _____

8. Are there existing structures on the site that must be demolished? ☐ ☐ ☐

   Comments: _____

   _____

9. If an existing structure must be demolished, does it present asbestos, lead paint, or unusual environmental problems? ☐ ☐ ☐

   Comments: _____

   _____

10. If the library is to be a branch of a public library system, are there overlapping service areas from other branches in the system? ☐ ☐ ☐

    Comments: _____

    _____

11. If the library is to be a branch of a college or university system, does the site provide ease of access for communication, transportation, and supply to the branch from other library service points? ☐ ☐ ☐

    Comments: _____

    _____

12. Will the use of the site for a library add aesthetic value or other amenities to the neighborhood? ☐ ☐ ☐

    Comments: _____

    _____

13. Are there liabilities or nuisance factors in adjacent properties and their activities? ☐ ☐ ☐

    Comments: _____

    _____

14. Will the use of the site for a library have any negative impact on the surrounding areas? ☐ ☐ ☐

    Comments: _____

    _____

15. Will the library fit in with the architectural style of neighboring buildings? ☐ ☐ ☐

    Comments: _____

    _____

16. Will the building work with the traffic flow of adjacent areas? ☐ ☐ ☐

    Comments: _____

    _____

	YES	NO	N/A

17. Will the site provide access options for people without cars? ☐ ☐ ☐

   Comments: _____

   _____

B. Community Planning Issues

1. Does the project respect the prevailing scale, pattern, and architectural character of the neighborhood where the library is planned? ☐ ☐ ☐

   Comments: _____

   _____

2. Does the site provide required and desirable utilities? These utilities may include water, sewage systems, electrical utilities, cable, Internet, and so on. ☐ ☐ ☐

   Comments: _____

   _____

3. Will the library site have "better utilities" (such as buried electric lines) and therefore be less prone to outages caused by storms? Also, it is more attractive not having telephone poles and wires strung through a neighborhood. ☐ ☐ ☐

   Comments: _____

   _____

4. Will the site have new or upgraded roads? ☐ ☐ ☐

   Comments: _____

   _____

5. Does the building respect the topography of the site and the surrounding area? ☐ ☐ ☐

   Comments: _____

   _____

6. Is the building placed on its site so it responds to its position on the block and to the placement of surrounding buildings? ☐ ☐ ☐

   Comments: _____

   _____

7. Does the front setback provide a pedestrian scale and enhance the street? ☐ ☐ ☐

   Comments: _____

   _____

8. In areas with varied front setbacks, is the building designed to act as a transition between adjacent buildings and to unify the overall streetscape? ☐ ☐ ☐

   Comments: _____

   _____

	YES	NO	N/A

9. Does the building provide landscaping in the front setback? ☐ ☐ ☐

Comments: _____

_____

10. Is the building articulated to minimize impacts on light to adjacent properties? ☐ ☐ ☐

Comments: _____

_____

11. Is the building articulated to minimize impacts on privacy to adjacent properties? ☐ ☐ ☐

Comments: _____

_____

12. Is the building's facade width compatible with those found on surrounding buildings? ☐ ☐ ☐

Comments: _____

_____

13. Are the building's proportions compatible with those found on surrounding buildings? ☐ ☐ ☐

Comments: _____

_____

14. Is the building's roofline compatible with those found on surrounding buildings? ☐ ☐ ☐

Comments: _____

_____

15. Does the building entrance enhance the connection between the public realm of the street and sidewalk and the private realm of the building? ☐ ☐ ☐

Comments: _____

_____

16. Are utility panels located so they are not visible on the front building wall or on the sidewalk? ☐ ☐ ☐

Comments: _____

_____

17. Are the design and placement of the garage/service entrance and doors compatible with the building and the surrounding area? ☐ ☐ ☐

Comments: _____

_____

18. Does the building provide loading and parking facilities in the rear that can be accessed through an alley or secondary street? ☐ ☐ ☐

Comments: _____

_____

19. Do the windows contribute to the architectural character of the building and the neighborhood? ☐ ☐ ☐

Comments: _____

_____

| | YES | NO | N/A |

20. Are the type, finish, and quality of the building's materials compatible with those used in the surrounding area? ☐ ☐ ☐

    Comments: _____

    _____

21. Are the building's exposed walls covered and finished with quality materials that are compatible with the front facade and adjacent buildings? ☐ ☐ ☐

    Comments: _____

    _____

22. Are the building's materials properly detailed and appropriately applied? ☐ ☐ ☐

    Comments: _____

    _____

23. Does the project utilize innovative materials and design that enrich the architectural character of the area adjacent to the library site? ☐ ☐ ☐

    Comments: _____

    _____

C. Location

1. Does the population being served consider the location of the site satisfactory and acceptable? ☐ ☐ ☐

   Comments: _____

   _____

2. Is the site accessible to all segments of the community being served? ☐ ☐ ☐

   Comments: _____

   _____

3. Is the site relatively close to the part of the community that is understood to be most active, and that will generate the most use? ☐ ☐ ☐

   Comments: _____

   _____

4. Is the site appropriate for the library given its function and clientele? ☐ ☐ ☐

   Comments: _____

   _____

5. Would library usage:
   a) Increase if another site was selected? ☐ ☐ ☐
   b) Decrease if another site was selected? ☐ ☐ ☐
   c) Stay the same if another site was selected? ☐ ☐ ☐

   Comments: _____

   _____

6. Will this location best meet the library objective of providing materials and services to the greatest number of people at the lowest cost?   ☐ ☐ ☐

   Comments: _____

   _____

7. Is the location in an area that is frequently visited by members of the community for daily activities such as going to class, shopping, working, and seeking out other services?   ☐ ☐ ☐

   Comments: _____

   _____

8. Is the site located near commercial, retail, cultural, and other activities within the community?   ☐ ☐ ☐

   Comments: _____

   _____

9. Does the proposed site present a safety issue for patrons and library staff?   ☐ ☐ ☐

   Comments: _____

   _____

10. Is the site served by public transportation?   ☐ ☐ ☐

    Comments: _____

    _____

D. Accessibility

1. Is the site easily accessible to those living in the area served?   ☐ ☐ ☐

   Comments: _____

   _____

2. For academic libraries, is the library located to best serve the interests and unique needs of the college or university? However, there are a number of factors to consider:

   a) Is the library located in the center of the campus in order to link parts of the campus together? Most academic libraries are considered the "heart" of the campus, and the preferred location might be in the center of the campus.   ☐ ☐ ☐

   b) Does the library serve a commuter population? Libraries that serve primarily a commuter population might want to locate the library building near the transportation hub.   ☐ ☐ ☐

   c) Is the library close to most classrooms? Because the library needs to serve the entire population of the campus, it might be located at a site convenient to most classrooms.   ☐ ☐ ☐

   d) Is the library system a departmental system? If departmental or special subject libraries exist, the library site and its relation to the other libraries might also be considered.   ☐ ☐ ☐

   Comments: _____

   _____

3. For school libraries, have the following location factors been considered?
   Is the library close to:

   a) The study hall, to permit easy entry and return of students? ☐ ☐ ☐

   b) The theater or auditorium, to provide good access to projection and taping
   equipment and to other graphic and audiovisual support services? ☐ ☐ ☐

   c) A computer laboratory, to permit access to additional computers and peripheral
   equipment? ☐ ☐ ☐

   d) The teacher workroom, to encourage teachers to use the center and to put
   equipment and resources within reach? ☐ ☐ ☐

   e) Adjacent roads and/or service drives, to permit easy delivery of materials and
   after-hours access? ☐ ☐ ☐

   Comments: _____

   _____

4. For public libraries, the location should be driven by the same factors that influence
   retail site selection. A good rule of thumb is to try to get the site on the opposite
   corner from McDonald's. ☐ ☐ ☐

   Comments: _____

   _____

5. Can the greatest number of potential customers easily reach the site? ☐ ☐ ☐

   Comments: _____

   _____

6. Are travel times from target population areas to the library acceptable? ☐ ☐ ☐

   Comments: _____

   _____

7. Have automobile traffic patterns near the library been considered? ☐ ☐ ☐

   Comments: _____

   _____

8. Is the site located on a busy highway that will require a separate street-type
   entrance or driveway? ☐ ☐ ☐

   Comments: _____

   _____

9. Is the site accessible to public transportation? ☐ ☐ ☐

   Comments: _____

   _____

10. Is bicycle access encouraged and facilitated? ☐ ☐ ☐

    Comments: _____

    _____

11. Are there sidewalks for pedestrian access? ☐ ☐ ☐

    Comments: _____

    _____

	YES	NO	N/A

12. Is the site conveniently accessible to private vehicle transportation? ☐ ☐ ☐

    Comments: _____

    _____

13. Does the entrance to the library provide adequate space and ease of accessibility to accommodate all arriving individuals and groups at all times? ☐ ☐ ☐

    Comments: _____

    _____

E. Size

1. Does the size of the site provide adequate space for current needs? ☐ ☐ ☐

   Comments: _____

   _____

2. Will the site provide room for future expansion and/or remodeling? ☐ ☐ ☐

   Comments: _____

   _____

3. Does the site include enough space for appropriate amenities such as green space and landscaping? ☐ ☐ ☐

   Comments: _____

   _____

4. Is the site large enough to accommodate on-site parking? ☐ ☐ ☐

   Comments: _____

   _____

5. Does the property contain easements that may influence the type of construction that may occur on the site? ☐ ☐ ☐

   Comments: _____

   _____

6. Does the property accommodate adequate setbacks to meet zoning and aesthetic considerations? ☐ ☐ ☐

   Comments: _____

   _____

7. Is the property configuration adequate for successful completion of the building project? ☐ ☐ ☐

   Comments: _____

   _____

8. Is there enough space on the property and/or adjacent to it for staging during construction? ☐ ☐ ☐

   Comments: _____

   _____

F. Environmental Issues

1. Has an environmental impact report been made for the proposed site?

   ☐ ☐ ☐

   Comments: _____

   _____

2. Is the site oriented so that it is possible to take advantage of solar energy and/or photovoltaic systems?

   ☐ ☐ ☐

   Comments: _____

   _____

3. Are complications likely to arise from the nature of the ground beneath the building?

   ☐ ☐ ☐

   Comments: _____

   _____

4. Does the site have adequate drainage?

   ☐ ☐ ☐

   Comments: _____

   _____

5. Is the site above the level of a 100-year floodplain?

   ☐ ☐ ☐

   Comments: _____

   _____

6. Has a subsurface probe been done to examine soil conditions, utilities, and other factors?

   ☐ ☐ ☐

   Comments: _____

   _____

7. Has the condition of the soil been tested to determine the stability of the site, or any underground site problems?

   ☐ ☐ ☐

   Comments: _____

   _____

8. Are there any hidden problems of geology, topography, archaeology, buried objects, or toxic waste?

   ☐ ☐ ☐

   Comments: _____

   _____

9. Does the site present issues with indigenous peoples and endangered species?

   ☐ ☐ ☐

   Comments: _____

   _____

10. Has the site been improved; that is, are curbs, gutters, water, sewers, and electricity available?

    ☐ ☐ ☐

    Comments: _____

    _____

11. Are there any natural or artificial barriers that either limit access or usability
of the site?

☐ ☐ ☐

Comments: _____

_____

12. Do neighboring facilities pose possible environmental or nuisance problems?

☐ ☐ ☐

Comments: _____

_____

13. If the site is sloped, are there possible advantages or disadvantages to the slope?

☐ ☐ ☐

Comments: _____

_____

Sustainable Design

A. LEED Certification

1. The U.S. Green Building Council is made up of tens of thousands of member organizations, chapters, and student and community volunteers that share the same vision of a sustainable built environment for all within the next generation. They are the developers of LEED and provide the standards and means for a project to become LEED-certified. ☐ ☐ ☐

   Comments: _____

2. LEED, or Leadership in Energy & Environmental Design, is a green building certification program that recognizes best-in-class building strategies and practices. To receive LEED certification, building projects satisfy prerequisites and earn points to achieve different levels of certification. Prerequisites and credits differ for each rating system, and teams choose the best fit for their project. ☐ ☐ ☐

   Comments: _____

3. The latest update to the LEED (LEED v4) "went live" at the 2013 Conference of the U.S. Green Building Association in Philadelphia. LEED v4 is covered extensively at www.usgbc.org/LEED. ☐ ☐ ☐

   Comments: _____

4. Advantages of the LEED v4 over previous versions include:

   a) A focus on materials that goes beyond how much is used to get a better understanding of what's in the materials we spec for our buildings and the effect those components have on human health and the environment ☐ ☐ ☐

   b) Takes a more performance-based approach to indoor environmental quality to ensure improved occupant comfort ☐ ☐ ☐

   c) Brings the benefits of smart grid thinking to the forefront with a credit that rewards projects for participating in demand response programs ☐ ☐ ☐

   d) Provides a clearer picture of water efficiency by evaluating total building water use ☐ ☐ ☐

   Comments: _____

   _____

5. Many of the checklist items below are adapted from the U.S. Green Building Council recommendations. ☐ ☐ ☐

   Comments: _____

   _____

B. Sustainable Sites

1. Is the new building or renovation designed and constructed in ways that preserve the natural outdoor environment and promote a healthful indoor habitat? ☐ ☐ ☐

   Comments: _____

   _____

2. Is the building project designed to avoid inflicting permanent adverse impact on the natural state of the air, land, and water by using resources and methods that minimize pollution and waste, and do not cause permanent damage to the earth, including erosion? ☐ ☐ ☐

   Comments: _____

   _____

3. Is the building designed to take the maximum advantage of passive and natural sources of heat, cooling, ventilation, and light? ☐ ☐ ☐

   Comments: _____

   _____

4. Are innovative strategies and technologies such as porous paving used to conserve water and reduce effluent and runoff, thus recharging the water table employed? ☐ ☐ ☐

   Comments: _____

   _____

5. Is there an erosion and sedimentation control plan for all construction activities associated with the project? ☐ ☐ ☐

   Comments: _____

   _____

	YES	NO	N/A

6. Is it possible to locate the library within 1/4 mile of an existing, or planned and funded, commuter rail, light rail, or subway station? ☐ ☐ ☐

Comments: _____

_____

7. Does the building provide secure bicycle racks and/or storage (within 200 yards of a building entrance) for at least 5 percent of building users? ☐ ☐ ☐

Comments: _____

_____

8. Is there a shower and are there changing facilities in the building, or within 200 yards of a building entrance? ☐ ☐ ☐

Comments: _____

_____

9. Is the project planned to reduce the need for individual automobiles, use alternative fuels, and encourage public and alternate modes of transportation such as bicycling and public transportation? ☐ ☐ ☐

Comments: _____

_____

10. Does the library provide preferred parking for 5 percent of the total vehicle parking capacity of the site, and at least one designated carpool drop-off area, for low-emitting and hybrid fuel-efficient vehicles? Note that some communities and educational institutions are designing buildings to discourage private auto access. ☐ ☐ ☐

Comments: _____

_____

11. If the library charges for parking, does it provide a discount of no less than 20 percent off parking pass rates for low-emitting and fuel-efficient vehicles (LEV/FEV)? ☐ ☐ ☐

Comments: _____

_____

12. Does the library provide alternative-fuel fueling stations (hydrogen, electric, etc.) for 3 percent of total parking capacity? ☐ ☐ ☐

Comments: _____

_____

13. Does the library provide users with access to an LEV/FEV sharing program? ☐ ☐ ☐

Comments: _____

_____

14. Is the building project designed to avoid inflicting permanent adverse impact on the natural state of the air, land, and water by using resources and methods that minimize pollution and waste, and do not cause permanent damage to the earth, including erosion? ☐ ☐ ☐

Comments: _____

_____

15. Is the building designed to take the maximum advantage of passive and natural sources of heat, cooling, ventilation, and light?   ☐ ☐ ☐

    Comments: _____

    _____

16. Is the library building planned for a sustainable site? Some inappropriate sites include:

    a) Prime farmland as defined by the U.S. Department of Agriculture.

    b) Previously undeveloped land whose elevation is lower than five feet above the elevation of the 100-year floodplain as defined by the Federal Emergency Management Agency (FEMA).   ☐ ☐ ☐

    c) Land that is specifically identified as habitat for any species on federal or state threatened or endangered lists.   ☐ ☐ ☐

    d) Within 100 feet of any wetlands as defined by the U.S. Code of Federal Regulations.   ☐ ☐ ☐

    e) Previously undeveloped land that is within fifty feet of a water body, defined as seas, lakes, rivers, streams, and tributaries which support or could support fish, recreation or industrial use, consistent with the terminology of the Clean Water Act.   ☐ ☐ ☐

    f) Land, which prior to acquisition for the project, was public parkland, unless land of equal or greater value as parkland is accepted in trade by the public landowner.   ☐ ☐ ☐

    Comments: _____

    _____

17. Is it possible for the library site to rehabilitate damaged sites where development is complicated by environmental contamination, reducing pressure on undeveloped land (Brownfield sites)? (Brownfield sites are sites which have been previously built on and which have been, or can be cleared for redevelopment.)   ☐ ☐ ☐

    Comments: _____

    _____

18. On Brownfield sites, is it possible on previously developed or graded sites to restore the site with native or adapted vegetation?   ☐ ☐ ☐

    Comments: _____

    _____

19. Do Greenfield sites limit all site disturbances to the least possible area beyond the building perimeter? (Greenfield sites are areas of land which have not previously been built on, and which might well be put to agricultural or amenity use.)   ☐ ☐ ☐

    Comments: _____

    _____

20. Is it possible to cover 50 percent of the site (excluding building footprint) or 20 percent of the entire site with native or adapted vegetation? ("Native or adapted vegetation" is defined as plants indigenous to a locality or cultivars of native plants that are adapted to the local climate and are not considered invasive species or noxious weeds.)   ☐ ☐ ☐

    Comments: _____

    _____

	YES	NO	N/A

21. Is the new building or renovation designed and constructed in ways that preserve the natural outdoor environment and promote a healthful indoor habitat? ☐ ☐ ☐

   Comments: _____

   _____

22. Is it possible to reduce the development footprint (defined as the total area of the building footprint, hardscape, access roads, and parking) and/or provide vegetated open space within the project boundary to exceed the local zoning's open space requirement for the site? ☐ ☐ ☐

   Comments: _____

   _____

23. Is it possible to implement a storm water management plan that reduces impervious cover, promotes infiltration, and captures and treats the storm water runoff from 90 percent of the average annual rainfall using acceptable best management practices (BMPs)? ☐ ☐ ☐

   Comments: _____

   _____

24. Will the post-development storm water peak discharge rate not exceed the pre-development rate? ☐ ☐ ☐

   Comments: _____

   _____

25. Can heat islands (thermal gradient differences between developed and undeveloped areas) be minimized through shade, paving materials, and an open grid pavement system in order to minimize the impact on microclimate and human and wildlife habitats? ☐ ☐ ☐

   Comments: _____

   _____

26. Can roofing materials and/or a vegetated room be used to reduce thermal heat islands? ☐ ☐ ☐

   Comments: _____

   _____

27. Can the light trespass from the building and site be minimized to reduce sky-glow to increase night sky access, improve nighttime visibility through glare reduction, and reduce development impact on nocturnal environments? ☐ ☐ ☐

   Comments: _____

   _____

28. Is the building site designed to take the maximum advantage of passive and natural sources of heat, cooling, ventilation, and light? ☐ ☐ ☐

   Comments: _____

   _____

29. Is it possible to make the library a more integrated part of the community by enabling the building and its grounds to be used for non-library events and functions? ☐ ☐ ☐

Comments: _____

_____

C. Water Efficiency

1. Is only captured rainwater, recycled wastewater, recycled gray water, or water treated and conveyed by a public agency specifically for non-potable uses employed for irrigation? ☐ ☐ ☐

Comments: _____

_____

2. Will a goal of capturing and treating 90 percent of runoff from annual rainfall be established? ☐ ☐ ☐

Comments: _____

_____

3. Are innovative strategies and technologies such as porous paving to conserve water and reduce effluent and runoff, thus recharging the water table, being employed? ☐ ☐ ☐

Comments: _____

_____

4. Has landscaping (native plants) been installed that does not require permanent irrigation systems? Temporary irrigation systems used for plant establishment should be removed within one year of installation. Native plants in a particular area are those that were growing naturally in the area before humans introduced plants from distant places. ☐ ☐ ☐

Comments: _____

_____

5. Has potable water use for building sewage conveyance been reduced through the use of water-conserving fixtures (water closets, urinals) or non-potable water (captured rainwater, recycled gray water, and on-site or municipally treated wastewater)? ☐ ☐ ☐

Comments: _____

_____

6. Are high-efficiency fixtures and dry fixtures such as composting toilet systems and non-water-using urinals used to reduce wastewater? ☐ ☐ ☐

Comments: _____

_____

7. Has reuse of storm water and gray water for non-potable applications such as toilet and urinal flushing, mechanical systems, and custodial uses been considered? ☐ ☐ ☐

Comments: _____

_____

8. Has refrigeration equipment using once-through cooling with potable water not been used? Not using this type of system increases water efficiency and reduces the burden on the municipal water supply and wastewater systems. ☐ ☐ ☐

Comments: _____

_____

D. Energy and Atmosphere

1. Are the building envelope and systems designed to maximize energy performance? ☐ ☐ ☐

Comments: _____

_____

2. Has a computer simulation model been used to assess the energy performance and to identify the most cost-effective energy efficiency measures? ☐ ☐ ☐

Comments: _____

_____

3. Have on-site renewable energy systems been employed to offset building energy costs? ☐ ☐ ☐

Comments: _____

_____

4. Can all or part of the energy needs of the building be met by green power? Green power is derived from solar, wind, geothermal, biomass, or low-impact hydro sources. ☐ ☐ ☐

Comments: _____

_____

5. Are the building's energy-related systems inspected, calibrated, and verified to determine that they perform according to the installed requirements, basis of design, and construction documents on a regular basis? ☐ ☐ ☐

Comments: _____

_____

6. Have refrigerants and HVAC&R (heating, ventilating, air-conditioning, and refrigerating) been selected that minimize or eliminate the emission of compounds that contribute to ozone depletion and global warming? ☐ ☐ ☐

Comments: _____

_____

7. Is there zero use of CFC-based refrigerants in the building HVAC&R systems? ☐ ☐ ☐

Comments: _____

_____

8. Are there fire suppression systems that contain ozone-depleting substances (CFCs, HCFCs, or Halons)? Substitutes might be a fire detection system and a zoned sprinkler system. ☐ ☐ ☐

Comments: _____

_____

9. Have photovoltaic cells been planned to help generate electricity? Photovoltaic cells convert light into electricity at the atomic level. There is usually a cost-benefit analysis that needs to be done to determine the economic benefits of photovoltaics. ☐ ☐ ☐

Comments: _____

_____

E. Materials

1. In planning for a new library building, has the reuse of existing buildings, including structure, envelope, and interior non-structural elements been considered? ☐ ☐ ☐

Comments: _____

_____

2. Has construction, demolition, and land-clearing debris been diverted from disposal in landfills and incinerators to recycling? Redirect recyclable recovered resources back to the manufacturing process. Redirect reusable materials to appropriate sites. ☐ ☐ ☐

Comments: _____

_____

3. Have opportunities to incorporate salvaged materials into building design been considered and potential material suppliers been researched? Salvaged materials such as beams and posts, flooring, paneling, doors and frames, cabinetry and furniture, and brick and decorative items should be considered. ☐ ☐ ☐

Comments: _____

_____

4. Have materials with recycled content been considered for use in new construction? ☐ ☐ ☐

Comments: _____

_____

5. Is the building constructed and operated using materials, methods, and mechanical and electrical systems that ensure a healthful indoor air quality, while avoiding contamination by carcinogens, volatile organic compounds, fungi, molds, bacteria, and other known toxins? ☐ ☐ ☐

Comments: _____

_____

6. Are copy rooms and similar spaces that emit possibly toxic substances equipped with their own dedicated air exhaust systems? ☐ ☐ ☐

Comments: _____

_____

7. Are particleboards that emit formaldehyde emissions prohibited in the building? ☐ ☐ ☐

Comments: _____

_____

	YES	NO	N/A

8. Are only solvent-free paints specified for the project? ☐ ☐ ☐

   Comments: _____

   _____

9. Are low-emitting, solvent-free adhesives specified for the project? ☐ ☐ ☐

   Comments: _____

   _____

10. Is furniture constructed without particleboards that emit formaldehyde? ☐ ☐ ☐

   Comments: _____

   _____

11. In evaluating the environmental performance of materials used in the building,
    have the following been considered in evaluating the materials?

   a) Energy efficient and with low embodied energy? ☐ ☐ ☐
   b) Made of renewable materials? ☐ ☐ ☐
   c) Made of post-consumer recycled materials? ☐ ☐ ☐
   d) Made of post-industrial recycled materials? ☐ ☐ ☐
   e) Made of certified wood? ☐ ☐ ☐
   f) Healthy for indoor air—low VOCs (volatile organic compounds)? ☐ ☐ ☐
   g) Healthy for the atmosphere—no CFCs or HCFCs used in manufacturing? ☐ ☐ ☐
   h) Nontoxic in use, production, or at end of useful life? ☐ ☐ ☐
   i) Made of salvaged materials? ☐ ☐ ☐
   j) Recyclable at end of useful life? ☐ ☐ ☐
   k) Simple to install without dangerous adhesives, and so on? ☐ ☐ ☐
   l) Made near to the building site—low transportation impacts? ☐ ☐ ☐
   m) Efficient/resourceful/reusable packaging? ☐ ☐ ☐

   Comments: _____

   _____

12. Is there a project goal for locally sourced materials, and have those materials and
    material suppliers that can achieve this goal been identified? ☐ ☐ ☐

   Comments: _____

   _____

13. Have rapidly renewable building materials and products (made from plants that
    are typically harvested within a ten-year cycle or shorter) been specified? Consider
    materials such as bamboo, wool, cotton insulation, agrifiber, linoleum, wheatboard,
    strawboard, and cork. ☐ ☐ ☐

   Comments: _____

   _____

14. Is there a location that provides an easily accessible area that serves the entire
    building and is dedicated to the collection and storage of non-hazardous materials
    for recycling, including (at a minimum) paper, corrugated cardboard, glass, plastics,
    and metals? ☐ ☐ ☐

   Comments: _____

   _____

F. Indoor Environmental Air Quality

1. Is smoking prohibited in the building?

   Comments: _____

   _____

   ☐ ☐ ☐

2. Is smoking prohibited outside the building?

   Comments: _____

   _____

   ☐ ☐ ☐

3. If outdoor smoking is allowed, are exterior designated smoking areas at least twenty-five feet away from entries, outdoor air intakes, and operable windows?

   Comments: _____

   _____

   ☐ ☐ ☐

4. Is there a permanent monitoring system that provides feedback on ventilation system performance to ensure that ventilation systems maintain design minimum ventilation requirements?

   Comments: _____

   _____

   ☐ ☐ ☐

5. Is there a building "flushout" after construction ends and prior to occupancy and after all interior finishes are installed? A building flushout is performed by supplying a total air volume of 14,000 cu. ft. of outdoor air per sq. ft. of floor area while maintaining an internal temperature of at least 60 degrees F and relative humidity no higher than 60 percent.

   Comments: _____

   _____

   ☐ ☐ ☐

6. Are carbon dioxide concentrations within all densely occupied spaces (those with a design occupant density greater than or equal to 25 people per 1000 sq. ft.) monitored? CO_2 monitoring locations shall be between 3 feet and 6 feet above the floor.

   Comments: _____

   _____

   ☐ ☐ ☐

7. Are any sources of chemicals that could be hazardous to occupants isolated? This may include separating copiers into spaces that can be properly ventilated so that the ozone from the copiers does not affect the entire library. This will also include keeping the pollutants from the streets, sidewalks, and parking lots out of the library by having effective walk-off mats at all main entryways.

   Comments: _____

   _____

   ☐ ☐ ☐

8. Are the HVAC system's outdoor air intakes located as high as possible above the ground and far enough away from the exhaust ducts to reduce the intake of ground-level air pollution (exhaust from traffic)?

   Comments: _____

   _____

   ☐ ☐ ☐

	YES	NO	N/A

9. Are air filters designed to be easy to access and clean and/or replace?   ☐ ☐ ☐

   Comments: _____

   _____

10. Are stainless-steel-strip bird guards installed over the horizontal rooftop outdoor air intakes to prevent birds from settling on the grating and polluting the shafts below?   ☐ ☐ ☐

   Comments: _____

   _____

11. Are air filters designed to be easy to access and clean and/or replace?   ☐ ☐ ☐

   Comments: _____

   _____

12. Has the exposed fiberglass (porous insulation) within the HVAC system been encapsulated to eliminate amplification sites for fungal and bacterial microorganisms?   ☐ ☐ ☐

   Comments: _____

   _____

13. Is the rate of ventilation with outdoor air at the rate of twenty-five cubic feet per minute?   ☐ ☐ ☐

   Comments: _____

   _____

G. Lighting and Day Lighting

1. Is lighting used for:
   a) Aesthetics to illuminate the exterior of the library and landscape?   ☐ ☐ ☐
   b) Security to illuminate the grounds near the library, driveway, and parking areas?   ☐ ☐ ☐
   c) Utility to illuminate the building, driveways and parking areas to help people navigate safely to and from the library?   ☐ ☐ ☐

   Comments: _____

   _____

2. Is day lighting used as a passive strategy in order to improve the indoor environmental quality of the library?   ☐ ☐ ☐

   Comments: _____

   _____

3. If day lighting is used, is there a daylight sensor to control the lights in day-lit spaces. Using photo-sensors in day-lit spaces to control dimmable ballasts will allow a system to work without being actively operated by occupants.   ☐ ☐ ☐

   Comments: _____

   _____

4. Is there a system to reduce the amount of time that lights are on? This can be accomplished by using dimmers and other lighting controls that turn off unneeded lights when a space is not occupied.   ☐ ☐ ☐

   Comments: _____

   _____

5. Is wattage reduced by replacing bulbs or entire fixtures with bulbs and fixtures that provide the same amount or greater amounts of light but with reduced electricity usage? Today, this can be accomplished most easily by replacing inefficient incandescent bulbs with incandescent/halogen bulbs or compact fluorescent bulbs consuming less energy in the turning off and on.   ☐ ☐ ☐

   Comments: _____

   _____

H. Roofs

1. Has a green roof been considered? A green roof system is an extension of the existing roof, and includes a high-quality waterproofing and root repellent system, a drainage system, filter cloth, and lightweight growing plants and vegetation. Green roof development involves the creation of "contained" green space on top of a human-made structure. This green space could be below, at, or above grade, but in all cases the plants are not planted in the "ground." Green roofs can provide a wide range of public and private benefits.   ☐ ☐ ☐

   Comments: _____

   _____

2. If a green roof is not used, has a cool roof been considered? Cool roofs use materials that reflect the sun's energy and help keep the indoor temperature down, reducing the heat load and requirements of the cooling system in the warmer months. Reflective or light-colored roofing materials are particularly recommended in hot and humid environments. Climate in specific areas of the country will determine whether to use a green or a cool roof.   ☐ ☐ ☐

   a) Cool roof materials take advantage of the reflective properties of light-colored materials. Dark materials absorb the energy from the sun and transfer this heat into your home. If your home has dark shingles, you may be running your air conditioner more than necessary, consuming extra energy and paying higher utility bills.   ☐ ☐ ☐

   Comments: _____

   _____

3. Has a roof with a photovoltaic (PV) energy system been considered? Photovoltaic systems convert sunlight directly into electricity.   ☐ ☐ ☐

   Comments: _____

   _____

I. Heating, Ventilating, and Air Conditioning (HVAC)

1. Has the HVAC been properly designed and installed? Proper heating, ventilating, and air conditioning are key to maintaining a comfortable, healthy, and productive work environment. Collectively, these systems account for approximately 40 percent of the electricity used in commercial buildings. Improved heating and cooling performance along with substantial energy savings can be achieved by implementing energy-efficiency measures. ☐ ☐ ☐

   Comments: _____

   _____

2. Have the cooling and heating load requirements been calculated? Accurate load requirement calculations (how much heating and cooling you actually need) will provide a system that meets the HVAC needs without adding excess capacity. Over-sizing equipment increases the capital cost at the time of the installation and the costs of operation of the equipment. ☐ ☐ ☐

   Comments: _____

   _____

3. Have possible ways to reduce load been considered? ☐ ☐ ☐

   a) Has reducing solar gain been considered? Solar gain should be reduced in warm climates and increased in cold climates. Solar gain may be reduced by using a cool roof and window tints. ☐ ☐ ☐

   Comments: _____

   _____

4. Have the most efficient office equipment and consumer products been specified to reduce heat gain? ☐ ☐ ☐

   Comments: _____

   _____

5. Have energy recovery ventilation systems been considered to reclaim waste energy? Using recovered energy from the exhaust air helps condition the incoming fresh air. ☐ ☐ ☐

   Comments: _____

   _____

6. Has supplemental dehumidification been considered in humid climates? By controlling humidity at your facility, you can increase occupant comfort and allow for further downsizing of equipment. ☐ ☐ ☐

   Comments: _____

   _____

7. In dry climates, have evaporative coolers been considered? These coolers use the evaporation of water to cool spaces, eliminating the need for energy-intensive compressors. ☐ ☐ ☐

   Comments: _____

   _____

8. Are programmable thermostats used to achieve peak energy savings? They may be controlled to reduce requirements when the building is closed. ☐ ☐ ☐

   Comments: _____

   _____

9. Are the thermostats only programmable by authorized library staff? ☐ ☐ ☐

   Comments: _____

   _____

10. Are premium variable speed drive (VSD) motors used on the HVAC equipment? VSD motors save energy on condenser and evaporator fans. ☐ ☐ ☐

    Comments: _____

    _____

11. Has radiant heating been considered for nonpublic areas such as warehouses and garages? Radiant heating warms objects instead of the air, and requires less fuel. ☐ ☐ ☐

    Comments: _____

    _____

5

General Exterior Considerations

A. Landscaping

1. Has the landscape design been considered early in the planning and design stage? ☐ ☐ ☐
    Comments: _____
    _____

2. Has a landscape architect been retained as one of the architect's sub-consultants? ☐ ☐ ☐
    Comments: _____
    _____

3. Does the landscape design enhance the overall design of the building? ☐ ☐ ☐
    Comments: _____
    _____

4. Does the landscaping complement and enhance the site and adjoining neighborhood? ☐ ☐ ☐
    Comments: _____
    _____

5. Is the landscaping visually satisfying and inviting? ☐ ☐ ☐
    Comments: _____
    _____

6. Is the landscaping design in harmony with the climatic zone of the library site? ☐ ☐ ☐
    Comments: _____
    _____

7. Has xeriscape been considered? Xeriscape is a combination of commonsense landscaping principles that save water while creating a lush and colorful landscape. It includes the following steps: ☐ ☐ ☐

   ☑ Plan and design the landscape plan.

   ☑ Create practical turf areas of manageable sizes, shapes, and appropriate grasses. Turf is used to enhance the landscaping and is used judiciously.

   ☑ Select and group plants appropriately. Different plants have different requirements when it comes to light, soil, and water.

   ☑ Use soil amendments like compost or manure as needed by the site and the type of plants used.

   ☑ Use mulches such as wood chips to reduce evaporation and to keep the soil cool.

   ☑ Irrigate efficiently with properly designed systems (including hose-end equipment) and by applying the right amount of water at the right time.

   ☑ Maintain the landscaping by mowing, weeding, pruning, and fertilizing properly.

   Comments: _____
   _____

8. Do the plants selected provide pleasing colors and textures throughout all seasons of the year? ☐ ☐ ☐

   Comments: _____
   _____

9. Is the landscaping designed from both an interior and exterior perspective to enhance the experience when one looks toward the library and looks out from the library? ☐ ☐ ☐

   Comments: _____
   _____

10. Is there an adequate amount of good soil? ☐ ☐ ☐

    Comments: _____
    _____

11. Is there adequate drainage? ☐ ☐ ☐

    Comments: _____
    _____

12. Are the plants selected appropriate to the amount of sun and/or shade they will receive? ☐ ☐ ☐

    Comments: _____
    _____

13. Do trees and shrubs enhance the building's energy and water conservation efforts? ☐ ☐ ☐

    Comments: _____
    _____

14. Has harvesting rainwater been considered? This can be as simple as placing a rain barrel under downspouts. The barrels will hold nutrient-rich rainwater for plant use. ☐ ☐ ☐

    Comments: _____
    _____

	YES	NO	N/A

15. Are the plants and shrubs selected not subject to damaging attacks by insects or disease? ☐ ☐ ☐

Comments: _____

_____

16. Can the landscaping be easily and inexpensively maintained? ☐ ☐ ☐

Comments: _____

_____

17. Is there an automatic irrigation system in place? ☐ ☐ ☐

Comments: _____

_____

18. Is the parking area landscaped in conformance with local codes and regulations? ☐ ☐ ☐

Comments: _____

_____

19. Is a local garden club or community organization willing to provide volunteer
gardening as a public service? ☐ ☐ ☐

Comments: _____

_____

B. Parking

1. Has the library made a policy decision on whether to provide parking for customers
and staff? Some libraries want to discourage the use of autos. ☐ ☐ ☐

Comments: _____

_____

2. Are there sufficient parking spaces for staff as well as customers during all service hours? ☐ ☐ ☐

Comments: _____

_____

3. Does the site provide adequate parking spaces to meet institutional and local
parking codes? ☐ ☐ ☐

Comments: _____

_____

4. Do handicapped parking spaces meet or exceed Americans with Disabilities Act
(ADA) regulations in both number and specifications? ☐ ☐ ☐

Comments: _____

_____

5. Is parking convenient to the library's entrances? ☐ ☐ ☐

Comments: _____

_____

	YES	NO	N/A

6. Is the parking area well lighted at night? ☐ ☐ ☐

   Comments: _____

   _____

7. Is there adequate parking for large cars and trucks? ☐ ☐ ☐

   Comments: _____

   _____

8. If there is a parking garage, is it close to the library's main entrance? ☐ ☐ ☐

   Comments: _____

   _____

9. Is the parking garage well identified from the street? ☐ ☐ ☐

   Comments: _____

   _____

10. Is the parking garage secure and well lighted at all times? ☐ ☐ ☐

   Comments: _____

   _____

11. Can cars easily get in and out of parking lots and/or structures? ☐ ☐ ☐

   Comments: _____

   _____

12. If the library has an employee recognition program, is there a designated parking space for "staffer of the month" very near the staff or receiving entrance? ☐ ☐ ☐

   Comments: _____

   _____

13. If there is a bookmobile, is parking convenient for staff to move materials on and off the vehicle? ☐ ☐ ☐

   Comments: _____

   _____

14. If there is a community room, is there adequate parking for the number of extra cars that will need to be parked? ☐ ☐ ☐

   Comments: _____

   _____

15. In northern climates, is there adequate room for snowplow access as well as snow stacking space? ☐ ☐ ☐

   Comments: _____

   _____

16. Does the library subsidize parking if free parking is not available? ☐ ☐ ☐

   Comments: _____

   _____

17. If the library parking is metered, does the library provide convenient coin-changing machines? ☐ ☐ ☐

Comments: _____

_____

18. If the library does not provide parking, is public parking available nearby? ☐ ☐ ☐

Comments: _____

_____

C. Building Exterior

1. Is the building aesthetically pleasing during the day and night? ☐ ☐ ☐

Comments: _____

_____

2. Is the fenestration arranged to take maximum advantage of natural light and the best views, while allowing use of floor and wall space inside the building? ☐ ☐ ☐

Comments: _____

_____

3. Will sunlight, glare, and excessive ultraviolet rays be controlled architecturally? ☐ ☐ ☐

Comments: _____

_____

4. Are all exterior architectural features and surfaces constructed of easily maintained materials? ☐ ☐ ☐

Comments: _____

_____

5. Do walls have a hard texture that is not easily scratched? ☐ ☐ ☐

Comments: _____

_____

6. Do walls have a graffiti-repellent coating? ☐ ☐ ☐

Comments: _____

_____

7. Do all exterior access walks and surfaces meet ADA requirements? ☐ ☐ ☐

Comments: _____

_____

8. Are all walkways and ramps leading into the building well lighted? ☐ ☐ ☐

Comments: _____

_____

9. In northern areas, do sidewalk lamps give off heat to help melt snow and ice? ☐ ☐ ☐

    Comments: _____

    _____

10. Are walkway surfaces stable and firm? ☐ ☐ ☐

    Comments: _____

    _____

11. Are walkway surfaces slip-resistant? ☐ ☐ ☐

    Comments: _____

    _____

12. Are stair steps uniform in height and width? ☐ ☐ ☐

    Comments: _____

    _____

13. Is there a separate staff entrance? ☐ ☐ ☐

    Comments: _____

    _____

14. Are public telephones available outside? ☐ ☐ ☐

    Comments: _____

    _____

15. Is there provision for storage of lawn mowers, snow blowers, and other outside
    equipment? ☐ ☐ ☐

    Comments: _____

    _____

16. Is there provision outside for vandal-proof faucets and electrical outlets? ☐ ☐ ☐

    Comments: _____

    _____

D. Roof

1. In northern areas, is the roof peaked to facilitate drainage? ☐ ☐ ☐

    Comments: _____

    _____

2. Are drainage systems on the roof adequate to carry off water from heavy downpours
    or melted snow? ☐ ☐ ☐

    Comments: _____

    _____

3. Are the roof and eaves areas well insulated to allow for maximum energy efficiency? ☐ ☐ ☐

    Comments: _____

    _____

	YES	NO	N/A

4. Is it possible to walk on the roof safely and without damaging it? ☐ ☐ ☐

Comments: _____

_____

5. Is the building's roof easily maintained? ☐ ☐ ☐

Comments: _____

_____

6. Are entrances and walkways protected from avalanches of water, snow, or ice accumulated on the roof? ☐ ☐ ☐

Comments: _____

_____

7. Do downspouts carry the water away from the building and sidewalks into storm drains? ☐ ☐ ☐

Comments: _____

_____

E. Bicycle Racks

1. Are bicycle racks clearly visible from the street and/or interior? ☐ ☐ ☐

Comments: _____

_____

2. Are bicycle racks convenient to the building entrances without being an obstacle for people entering the library? ☐ ☐ ☐

Comments: _____

_____

3. Are bicycle racks equipped with locks? ☐ ☐ ☐

Comments: _____

_____

4. Are the bicycle racks in a well-lighted area? ☐ ☐ ☐

Comments: _____

_____

F. Flagpole

1. Is there a flagpole outside the building? ☐ ☐ ☐

Comments: _____

_____

2. Is it a ground-set, wall-mounted, or roof-mounted pole? ☐ ☐ ☐

Comments: _____

_____

3. Is there a self-storing flagpole shaft? ☐ ☐ ☐

   Comments: _____

   _____

4. Can the flag be raised, lowered, and drawn into the pole either manually or electrically? ☐ ☐ ☐

   Comments: _____

   _____

5. Is the flagpole safe from vandalism? ☐ ☐ ☐

   Comments: _____

   _____

6. If the flag is to be flown at night, is it adequately illuminated? ☐ ☐ ☐

   Comments: _____

   _____

G. Exterior Signage

1. Is signage incorporated into the preliminary design of the site, parking, and building? ☐ ☐ ☐

   Comments: _____

   _____

2. Does signage comply with the ADA? ☐ ☐ ☐

   Comments: _____

   _____

3. Is the standard international symbol for libraries displayed? ☐ ☐ ☐

   Comments: _____

   _____

4. Is there a large, exterior, well-lit sign identifying the library? ☐ ☐ ☐

   Comments: _____

   _____

5. Is the exterior sign clearly visible from passing cars during the day and night? ☐ ☐ ☐

   Comments: _____

   _____

6. Does the sign(s) have space for advertising of library events, holiday hours, and so on? ☐ ☐ ☐

   Comments: _____

   _____

7. Are the library's hours of service prominently displayed on a large, well-lit sign at the entrance along with an OPEN/CLOSED sign? ☐ ☐ ☐

   Comments: _____

   _____

	YES	NO	N/A

8. Do the colors of the letters contrast with the color of the sign and complement the outside of the building? ☐ ☐ ☐

   Comments: _____

   _____

9. Are signs attached to the wall adjacent to the latch side of the door? ☐ ☐ ☐

   Comments: _____

   _____

10. Would a map, directory, or graphic be more appropriate than a sign? ☐ ☐ ☐

    Comments: _____

    _____

11. Do pictorial signs have verbal descriptions placed below the picture? ☐ ☐ ☐

    Comments: _____

    _____

12. Are the letters in sans serif or simple serif? ☐ ☐ ☐

    Comments: _____

    _____

13. Do signs have a non-glare finish? ☐ ☐ ☐

    Comments: _____

    _____

14. When selecting sign size, have background and distance been considered? ☐ ☐ ☐

    Comments: _____

    _____

15. Is sign size 1 inch for every 50 feet of visibility and a minimum of 3 inches? ☐ ☐ ☐

    Comments: _____

    _____

16. Has negative phrasing been avoided in signage? ☐ ☐ ☐

    Comments: _____

    _____

17. Are the signs durable and can they be easily and cost-effectively replaced? ☐ ☐ ☐

    Comments: _____

    _____

18. Are signs read horizontally and not vertically? ☐ ☐ ☐

    Comments: _____

    _____

19. If there is an arrow to indicate direction, is it separate from the lettered sign so that it can be changed if necessary? ☐ ☐ ☐

    Comments: _____

    _____

H. Loading Docks and Delivery

1. Is the loading dock and/or delivery area clearly marked and easily accessible from the street?

    ☐   ☐   ☐

    Comments: _____

    _____

2. Is there generous space for easy truck turnaround?

    ☐   ☐   ☐

    Comments: _____

    _____

3. Is there a buzzer and/or internal telephone at or near the delivery entrance?

    ☐   ☐   ☐

    Comments: _____

    _____

4. If there is no loading dock, is parking for delivery vehicles located close to the exit nearest the delivery or workroom?

    ☐   ☐   ☐

    Comments: _____

    _____

5. Is the loading dock located away from the primary work and public areas so that noise and fumes do not disturb staff or users?

    ☐   ☐   ☐

    Comments: _____

    _____

6. Is there a ramp provided from the loading dock down to the truck parking area to facilitate deliveries from small trucks and vans? This ramp should have a maximum slope of 1:12 and comply with UFAS/ADA Accessibility Guidelines ensuring that it may be easily maneuverable for deliveries on carts and dollies

    ☐   ☐   ☐

    Comments: _____

    _____

7. Are the loading docks located for easy access by service vehicles and separated from public entrances to the building, and public spaces?

    ☐   ☐   ☐

    Comments: _____

    _____

8. Are loading docks convenient to freight elevators so that service is segregated from the main passenger elevator lobbies and public corridors? The service route from the elevator should accommodate the transport of large items.

    ☐   ☐   ☐

    Comments: _____

    _____

| | YES | NO | N/A |

9. Are loading docks designed to accommodate vehicles used to deliver or pick up materials from the library? If the bed height of vans and trucks varies more than 18 inches, at least one loading berth should be equipped with a dock leveler. Typical docks are built 55 inches above grade level to accommodate most trucks. ☐ ☐ ☐

Comments: _____
_____

10. Does the loading dock have a minimum overhead clearance of fourteen feet? ☐ ☐ ☐

Comments: _____
_____

11. Is each truck position equipped with adjustable lighting fixtures for the illumination of the interior of trailers? ☐ ☐ ☐

Comments: _____
_____

12. Are loading docks protected with edge guards and dock bumpers? ☐ ☐ ☐

Comments: _____
_____

13. Are easy-access overhead coiling doors used for the loading docks? These doors should be able to close completely and lock after the loading docks are closed. At least one well-lit personnel door should be provided in addition to the overhead doors. ☐ ☐ ☐

Comments: _____
_____

14. Are noise mitigation strategies used? Noise reductions in the dock and noise transmission out of the dock are different design considerations. Mass and limpness/flexibility are two desirable attributes for a sound transmission barrier. Unpainted heavy masonry walls provide mass. Absorptive acoustical surfacing will reduce the noise level in the dock but will have little effect on the transmission outside it. Noise levels in the dock should be moderated to promote communication among users. ☐ ☐ ☐

Comments: _____
_____

15. Is resilient flooring used in offices adjacent to the loading docks? ☐ ☐ ☐

Comments: _____
_____

16. Is the truck's approach to the dock at grade or sloped away from the loading dock to prevent the collection of storm water near the dock? ☐ ☐ ☐

Comments: _____
_____

17. Are loading docks covered at least four feet beyond the edge of the platform over the loading berth to protect users and goods being unloaded?

☐ ☐ ☐

Comments: _____

_____

18. Is a staging area inside the building provided adjacent to the loading dock? It must be protected from the weather.

☐ ☐ ☐

Comments: _____

_____

19. Is a dock manager's room or booth located so the manager can keep the entire dock area in view and control the entrance and exit from the building? The flow of circulation into the dock should pass this control point, and access should be restricted to authorized personnel. Security cameras may serve as a backup.

☐ ☐ ☐

Comments: _____

_____

20. Are loading docks designed not to interfere with emergency egress routes from the building?

☐ ☐ ☐

Comments: _____

_____

21. Can examination of deliveries take place in the staging area and be monitored by the truck driver and a shipping/receiving clerk? Shipping and receiving logs should be kept.

☐ ☐ ☐

Comments: _____

_____

22. Is the delivery area a separate room?

☐ ☐ ☐

Comments: _____

_____

23. Are there two separate counters/tables in the delivery area so that delivery staff can distinguish between outgoing and incoming packages?

☐ ☐ ☐

Comments: _____

_____

24. Do the counters/tables have enough length and breadth to provide sufficient space for peak loading times?

☐ ☐ ☐

Comments: _____

_____

25. Are the counters/tables at a comfortable height so as to avoid physical injury from lifting?

☐ ☐ ☐

Comments: _____

_____

I. Outdoor Trash Enclosures

1. Is there provision for the temporary storage and pickup of trash? ☐ ☐ ☐

   Comments: _____

   _____

2. Is there a system in place to recycle paper, plastics, and so on? ☐ ☐ ☐

   Comments: _____

   _____

3. Are there separate containers for recyclables? ☐ ☐ ☐

   Comments: _____

   _____

4. Is the trash area secure from "dumpster divers"? ☐ ☐ ☐

   Comments: _____

   _____

5. Is the trash area isolated from contact with storm water flows originating outside
   the storage area? ☐ ☐ ☐

   Comments: _____

   _____

6. Are trash areas surrounded with a barrier sufficient to prevent all trash from being
   transported out of the storage area, except during collection? ☐ ☐ ☐

   Comments: _____

   _____

J. Outdoor Book and Media Returns

1. Is there an after-hours book and media return? ☐ ☐ ☐

   Comments: _____

   _____

2. Does the return have separate slots for books and media? ☐ ☐ ☐

   Comments: _____

   _____

3. Does the return meet ADA requirements? ☐ ☐ ☐

   Comments: _____

   _____

4. Is the return area well lighted and secure? ☐ ☐ ☐

   Comments: _____

   _____

	YES	NO	N/A

5. Is the return sheltered from the weather and small creatures? ☐ ☐ ☐

    Comments: _____

    _____

6. Is the return part of the building and accessible from the inside rather than separate from the building? ☐ ☐ ☐

    Comments: _____

    _____

7. Is the return fire-retardant? ☐ ☐ ☐

    Comments: _____

    _____

8. Does the return area have a smoke detector? ☐ ☐ ☐

    Comments: _____

    _____

9. Is the return visible to patrons in automobiles? ☐ ☐ ☐

    Comments: _____

    _____

10. Is the return accessible from an automobile? ☐ ☐ ☐

    Comments: _____

    _____

11. Is the return designed so that it will not damage library materials as it is used? ☐ ☐ ☐

    Comments: _____

    _____

12. Does the door on the book return lock when the cart is full to prevent the cart overflowing? ☐ ☐ ☐

    Comments: _____

    _____

13. Is there a locking device on outside returns? ☐ ☐ ☐

    Comments: _____

    _____

14. Do outside returns accommodate both walk-up and drive-up access through two deposit openings? ☐ ☐ ☐

    Comments: _____

    _____

Interior Organization of Library Buildings

YES | NO | N/A

A. Entrance

1. For security purposes, is there only one public entrance/exit?

   Comments: _____

   _____

2. Is the staff entrance secured from unauthorized use and well lighted?

   Comments: _____

   _____

3. Is the building's entrance easily identifiable to pedestrians as well as people in cars?

   Comments: _____

   _____

4. Is the route from the public transportation stop to the entrance easily accessible?

   Comments: _____

   _____

5. Are all building entrances sheltered from the weather and well lighted?

   Comments: _____

   _____

6. Is a floor covering or system provided near the entrance that allows for removal of debris from users' shoes as they walk into the building?

   Comments: _____

   _____

7. Is there a floor drain provided for exterior rain and snow removal at the entrance to the building?　☐　☐　☐

   Comments: _____

   _____

8. Are there trash and cigarette receptacles near each of the entrances?　☐　☐　☐

   Comments: _____

   _____

9. Are the outside telephones well lighted at night and easily visible?　☐　☐　☐

   Comments: _____

   _____

10. If the library is at an intersection, is there a main entrance at or near a corner that will serve both streets?　☐　☐　☐

    Comments: _____

    _____

11. Is there outside seating available?　☐　☐　☐

    Comments: _____

    _____

12. Is there a double-door vestibule to prevent drafts and heat and/or air conditioning losses?　☐　☐　☐

    Comments: _____

    _____

13. Is the hardware for the entrance doors durable and sturdy enough to withstand heavy use?　☐　☐　☐

    Comments: _____

    _____

14. Are entrance doors easy to open and close?　☐　☐　☐

    Comments: _____

    _____

15. Are entrance doors relatively quiet?　☐　☐　☐

    Comments: _____

    _____

16. If glass is used near the entry, is it safety glass?　☐　☐　☐

    Comments: _____

    _____

17. Are all public-service elements of the building easily located from the entrance?　☐　☐　☐

    Comments: _____

    _____

	YES	NO	N/A

18. Can the book security system be installed without surface-mounted wiring or carpet runners? ☐ ☐ ☐

    Comments: _____

    _____

19. If a metal studding system is used in framing the building, are wood studs used adjacent to the area where the book security system is installed to prevent interference? ☐ ☐ ☐

    Comments: _____

    _____

20. Do signs, lighting, color, and furnishings identify the various areas within the interior? ☐ ☐ ☐

    Comments: _____

    _____

21. Do the areas listed below stand out when one enters the building:

    a) Circulation? ☐ ☐ ☐
    b) Reference/information? ☐ ☐ ☐
    c) Catalog? ☐ ☐ ☐
    d) Books/audiovisual? ☐ ☐ ☐
    e) Children/adults/young adults? ☐ ☐ ☐

    Comments: _____

    _____

22. Are furniture and equipment used to promote, merchandise, and display some parts of the book and media collections of the library? ☐ ☐ ☐

    Comments: _____

    _____

23. Is there space near the entrance for:

    a) Electronic public information kiosks? ☐ ☐ ☐
    b) Public bulletin boards? ☐ ☐ ☐
    c) Display cases? ☐ ☐ ☐
    d) Pamphlet racks? ☐ ☐ ☐
    e) Public/community announcements bulletin boards? ☐ ☐ ☐
    f) Public telephones? ☐ ☐ ☐
    g) Vending machines? ☐ ☐ ☐
    h) Book donation drop? ☐ ☐ ☐
    i) Lobby seating? ☐ ☐ ☐

    Comments: _____

    _____

24. Does there appear to be good traffic flow throughout the interior? ☐ ☐ ☐

    Comments: _____

    _____

B. Circulation Desk Facilities

1. Is the circulation area located near the library's entrance? ☐ ☐ ☐

   Comments: _____

   _____

2. Is the circulation area clearly visible and identifiable from the library's entrance? ☐ ☐ ☐

   Comments: _____

   _____

3. Is there enough space between the circulation and security equipment to prevent
   one system from interfering with the electrical and physical operation of the other? ☐ ☐ ☐

   Comments: _____

   _____

4. Are the following functions easily identified and located by library users:

   a) Checkout? ☐ ☐ ☐
   b) Self- or express checkout (if available)? ☐ ☐ ☐
   c) Returns? ☐ ☐ ☐
   d) Issuance of new library cards? ☐ ☐ ☐
   e) Information/inquiry? ☐ ☐ ☐
   f) Reserve/holds? ☐ ☐ ☐
   g) Interlibrary loan? ☐ ☐ ☐
   h) Other functions? ☐ ☐ ☐

   Comments: _____

   _____

5. Are queuing provisions made for a smooth traffic flow for entering and leaving the
   building without obstacles created by checkout lines during peak periods? ☐ ☐ ☐

   Comments: _____

   _____

6. Will checkout lines be long enough to require stanchions and roping? ☐ ☐ ☐

   Comments: _____

   _____

7. Does the circulation desk accommodate:

   a) Computer checkout terminals? ☐ ☐ ☐
   b) Self-checkout terminals? ☐ ☐ ☐
   c) Terminal screens that are visible to customers? ☐ ☐ ☐
   d) Telephones? ☐ ☐ ☐
   e) Answering machines? ☐ ☐ ☐
   f) Cash registers and/or cash drawers? ☐ ☐ ☐
   g) Lost and found items? ☐ ☐ ☐

   Comments: _____

   _____

	YES	NO	N/A

8. Are there sufficient sorting shelves and trucks for holding returned materials? ☐ ☐ ☐

   Comments: _____

   _____

9. Are the shelves and trucks easily accessible and clearly arranged? ☐ ☐ ☐

   Comments: _____

   _____

10. Can the sorting shelves accommodate all sizes of returned materials? ☐ ☐ ☐

    Comments: _____

    _____

11. Is there an interior book drop and can it be easily cleared? ☐ ☐ ☐

    Comments: _____

    _____

12. Is there adequate workspace for staff? ☐ ☐ ☐

    Comments: _____

    _____

13. Is there toe space and knee space incorporated into the circulation counter for staff comfort and convenience? ☐ ☐ ☐

    Comments: _____

    _____

14. Is the circulation desk the appropriate height for adults, children, and disabled customers? This will probably require different heights for sections of the desk in order to make the desk accessible and easy to use for all. Staff may also have preferences for workstations for standing or sitting. Desks to accommodate people with disabilities also have unique requirements, with knee space for people in chairs. ☐ ☐ ☐

    Comments: _____

    _____

15. Is the desk designed for a logical workflow based on the circulation system employed by the library? ☐ ☐ ☐

    Comments: _____

    _____

16. If there is a materials security system, is there space for the sensitizing and desensitizing equipment? ☐ ☐ ☐

    Comments: _____

    _____

17. Is there adequate space for book trucks to move about and through the circulation area? ☐ ☐ ☐

    Comments: _____

    _____

18. Are sorting shelves and trucks easily accessible from the return portions of the desk? ☐ ☐ ☐

Comments: _____

_____

19. Is the electrical wiring and cabling required to support equipment in the desk out of public view? ☐ ☐ ☐

Comments: _____

_____

20. Are there back panels on the computers to screen them from the public and protect against vandalism? ☐ ☐ ☐

Comments: _____

_____

21. Are the electrical wiring and cabling easily accessible by staff? ☐ ☐ ☐

Comments: _____

_____

22. Is the circulation desk designed to accommodate changing the location and size of electrical equipment in the future? This requires flexibility. ☐ ☐ ☐

Comments: _____

_____

23. Are keyboards ergonomically designed? ☐ ☐ ☐

Comments: _____

_____

24. Is the monitor screen visible to the customers? ☐ ☐ ☐

Comments: _____

_____

25. Is the top of the desk covered with a material that does not get damaged when heavy materials and equipment are dragged across or dropped upon it? ☐ ☐ ☐

Comments: _____

_____

26. Does the surface of the desk provide a good surface for writing? ☐ ☐ ☐

Comments: _____

_____

27. Can the desk surface be cleaned easily? ☐ ☐ ☐

Comments: _____

_____

28. Are the height and width of the circulation desk appropriate for the various work functions taking place? ☐ ☐ ☐

Comments: _____

_____

	YES	NO	N/A

29. Is the flooring material adjacent to the circulation counter of a type that will minimize the noise of book trucks? ☐ ☐ ☐

    Comments: _____

    _____

30. Is there shock-absorbent flooring next to the staff side of the circulation desk? ☐ ☐ ☐

    Comments: _____

    _____

31. Is the floor adjacent to the circulation counter easily maintained and safe during wet weather? ☐ ☐ ☐

    Comments: _____

    _____

32. Is the circulation desk a modular design so that modules may be interchanged as need arises? ☐ ☐ ☐

    Comments: _____

    _____

33. Is the desk designed to handle the necessary equipment with hidden, yet easily accessible, wiring and cable? ☐ ☐ ☐

    Comments: _____

    _____

34. Is there room to expand the desk as circulation of materials increases? ☐ ☐ ☐

    Comments: _____

    _____

35. Are circulation staff offices located near the circulation area, and easily accessible by the public? ☐ ☐ ☐

    Comments: _____

    _____

C. Reference Facilities

1. Is the reference desk clearly identified and conveniently located? ☐ ☐ ☐

    Comments: _____

    _____

2. Is the reference desk the appropriate height for adults, children, and disabled patrons? ☐ ☐ ☐

    Comments: _____

    _____

3. Is the reference area arranged in such a manner that librarians are visibly approachable? ☐ ☐ ☐

    Comments: _____

    _____

4. Is the reference desk located where staff can identify by sight those customers having difficulty finding reference materials?   ☐ ☐ ☐

    Comments: _____

    _____

5. Is there seating for customer/staff consultation?   ☐ ☐ ☐

    Comments: _____

    _____

6. Does the library employ a "tiered" reference system? This might include:

    a) Inquiry or informational desk at the main entrance to greet library users and direct them to the location they require?   ☐ ☐ ☐
    b) Main reference desk employing reference specialists to answer most of the library user's questions?   ☐ ☐ ☐
    c) Reference or bibliographic consultation stations adjacent to subject collections serviced by bibliographic specialists?   ☐ ☐ ☐

    Comments: _____

    _____

7. Can reference librarians easily get out from behind the desk to help customers?   ☐ ☐ ☐

    Comments: _____

    _____

8. Are reference collections, including ready reference materials, conveniently located and identified?   ☐ ☐ ☐

    Comments: _____

    _____

9. Are photocopiers close to the reference materials?   ☐ ☐ ☐

    Comments: _____

    _____

10. Are materials and equipment requiring staff assistance grouped close to the reference service desk?   ☐ ☐ ☐

    Comments: _____

    _____

11. Is there a terminal on the reference desk that can perform circulation functions as well as database searching functions?   ☐ ☐ ☐

    Comments: _____

    _____

12. Is the public access catalog accessible from all parts of the reference collection?   ☐ ☐ ☐

    Comments: _____

    _____

	YES	NO	N/A

13. Are catalog terminals well distributed in the reference area? ☐ ☐ ☐

   Comments: _____

   _____

14. Do the reference staffs have adequate workspace at their public service desk? ☐ ☐ ☐

   Comments: _____

   _____

15. Does the reference desk have a cordless or cellular phone in order to do more efficient
   interviews with telephone service customers while performing shelf checks? ☐ ☐ ☐

   Comments: _____

   _____

16. Does the telephone system have a multiline capacity? ☐ ☐ ☐

   Comments: _____

   _____

17. Are adequate space, appropriate lighting, and acoustics allowed for the following
   equipment and its use:

   a) Electronic workstations? ☐ ☐ ☐
   b) Audiovisual equipment? ☐ ☐ ☐
   c) Photocopiers? ☐ ☐ ☐
   d) Microform equipment? ☐ ☐ ☐
   e) Other? ☐ ☐ ☐

   Comments: _____

   _____

18. Is adequate space allowed for customer use of reference materials? ☐ ☐ ☐

   Comments: _____

   _____

19. Does the reference area provide separate or acoustically isolated spaces for the
   following services:

   a) Interlibrary loan? ☐ ☐ ☐
   b) Database searches? ☐ ☐ ☐
   c) General information? ☐ ☐ ☐
   d) Customer interviews? ☐ ☐ ☐
   e) Telephone and/or electronic reference service? ☐ ☐ ☐

   Comments: _____

   _____

20. If the following materials are included in the reference collection, is adequate space
   allowed for their use, including the staff, equipment, shelving, and furniture they require:

   a) General reference materials? ☐ ☐ ☐
   b) Electronic databases? ☐ ☐ ☐

c) Newspapers? ☐ ☐ ☐
d) Newspaper clippings? ☐ ☐ ☐
e) Periodicals? ☐ ☐ ☐
f) Indexes and abstracts? ☐ ☐ ☐
g) Annual reports? ☐ ☐ ☐
h) Microforms? ☐ ☐ ☐
i) Rare books? ☐ ☐ ☐
j) Bibliographies? ☐ ☐ ☐
k) Unabridged dictionaries? ☐ ☐ ☐
l) Government publications? ☐ ☐ ☐
m) Vertical files? ☐ ☐ ☐
n) College catalogs and career information? ☐ ☐ ☐
o) Ready reference? ☐ ☐ ☐
p) Reserves? ☐ ☐ ☐
q) City directories? ☐ ☐ ☐
r) Archives? ☐ ☐ ☐
s) Telephone directories? ☐ ☐ ☐
t) Genealogy resources? ☐ ☐ ☐
u) Maps and atlases? ☐ ☐ ☐
v) Audiovisual materials? ☐ ☐ ☐
w) Tax forms? ☐ ☐ ☐
x) General information flyers? ☐ ☐ ☐
y) Miscellaneous library and public information? ☐ ☐ ☐
z) Other? ☐ ☐ ☐

Comments: _____

_____

21. Are reference staff offices located near the reference area? ☐ ☐ ☐

Comments: _____

_____

22. If areas of limited or closed access exist, is adequate space allocated for:

a) Staffing? ☐ ☐ ☐
b) Expansion? ☐ ☐ ☐
c) Security? ☐ ☐ ☐

Comments: _____

_____

23. Does the reference staff have adequate workspace at their public service desks? ☐ ☐ ☐

Comments: _____

_____

24. Can the reference area be expanded for additional staff, equipment, shelving, and furniture? ☐ ☐ ☐

Comments: _____

_____

D. Information Commons

1. Does the library have an information commons? The information commons supports and enhances student learning and research by providing state-of-the-art technology and resources in an academic environment. It is a place for students, staff, and faculty to interact, get technology support and research assistance, attend technology and research workshops or classes, and work in groups together or individually on course assignments. ☐ ☐ ☐

   Comments: _____

   _____

2. Are there information and technology specialists who work side by side to serve students, faculty, and staff? ☐ ☐ ☐

   Comments: _____

   _____

3. Are there reference and/or service desk(s) in the information commons? ☐ ☐ ☐

   Comments: _____

   _____

4. Does the information commons offer reference services such as:
   a) Research for course assignments and personal interest? ☐ ☐ ☐
   b) Identifying and locating library materials in all formats, including print and electronic? ☐ ☐ ☐
   c) Identifying primary and secondary information sources? ☐ ☐ ☐
   d) Database searching (e.g., using Boolean operators, controlled vocabulary, etc.)? ☐ ☐ ☐
   e) Effective use of the World Wide Web, including how to evaluate websites? ☐ ☐ ☐

   Comments: _____

   _____

5. Are there information technology courses offered by library and/or information technology (IT) staff? ☐ ☐ ☐

   Comments: _____

   _____

6. Is there workspace for "blended librarians" and "roving librarians"? ☐ ☐ ☐

   Comments: _____

   _____

7. Does the information commons offer a variety of workstations including computers (MAC and PC), printers, scanners, and faxes? ☐ ☐ ☐

   Comments: _____

   _____

8. Is there a mix of individual and group workstations? ☐ ☐ ☐

   Comments: _____

   _____

	YES	NO	N/A

9. Is there a replacement cycle for the hardware? ☐ ☐ ☐

   Comments: _____

   _____

10. Does the information commons provide access to the library's collection, the Internet,
    and commercial databases? ☐ ☐ ☐

    Comments: _____

    _____

11. Are there multimedia work stations that allow users to:

    a) Digitize video from VHS, mini-DV, or camcorder? ☐ ☐ ☐
    b) Digitize analog audio from cassette tape? ☐ ☐ ☐
    c) Edit video and analog media? ☐ ☐ ☐
    d) Export digital audio and video to tape? ☐ ☐ ☐
    e) Scan images and text? ☐ ☐ ☐

    Comments: _____

    _____

12. Is there an adaptive technology center designed to assist people with disabilities?
    Does the center include:

    a) High-speed scanning? ☐ ☐ ☐
    b) Other alternate media: Braille, tactile graphics, MP3 audio books, and so on? ☐ ☐ ☐
    c) Hardware and software loaner programs? ☐ ☐ ☐
    d) Web accessibility consulting? ☐ ☐ ☐
    e) Does the center offer Kurzweil scan and read software that makes printed or
       electronic text accessible to people who are blind or visually impaired? ☐ ☐ ☐

    Comments: _____

    _____

13. Are there policies for use of the information commons, including:

    a) Curtailing disruptive behavior by making people exhibiting that behavior to leave? ☐ ☐ ☐
    b) Restarting computers after a period of time if a workstation is unattended and
       there are customers waiting to use the workstation? ☐ ☐ ☐
    c) Prohibiting sleeping, offensive bodily hygiene, or behavior that results in
       complaints or threatens to damage library property, disruptive behavior,
       food/beverage distribution, soliciting, roughhousing, pets, rollerblades/skates,
       bicycles, and so on, for example? ☐ ☐ ☐
    d) Requesting that cell phones be turned to "vibrate" and asking people to only
       use phones in areas that will not disturb other users? ☐ ☐ ☐

    Comments: _____

    _____

14. Does the information commons have group study rooms with computers, display
    screens, and videotaping systems? ☐ ☐ ☐

    Comments: _____

    _____

E. Multimedia Facilities

1. Does the area provide an opportunity to market multimedia materials and services to users through visual and sound displays? ☐ ☐ ☐

   Comments: _____
   _____

2. Does the media room have a separate, independent heating/cooling system that can be regulated to control the temperature and humidity? ☐ ☐ ☐

   Comments: _____
   _____

3. Is there special humidifying/dehumidifying equipment to maintain a 60 percent relative humidity? ☐ ☐ ☐

   Comments: _____
   _____

4. Do air conditioning units have electrostatic filters? ☐ ☐ ☐

   Comments: _____
   _____

5. Are supply and return air vents located high on the walls or in the ceiling with air velocities low enough to prevent draft problems with paper, hair, or clothing? ☐ ☐ ☐

   Comments: _____
   _____

6. Can windows be opened to provide ventilation in case the HVAC system breaks down? ☐ ☐ ☐

   Comments: _____
   _____

7. Is there sufficient acoustical treatment to prevent external noise sources from interfering with listening to media? ☐ ☐ ☐

   Comments: _____
   _____

8. Has the following equipment been considered for placement in multimedia areas:

   a) Electronic carrels with built-in playback equipment? ☐ ☐ ☐
   b) Secured and locked storage cabinets for equipment such as videotape recorders, cassette players, overhead movie and slide projectors, and so on? ☐ ☐ ☐
   c) Electronic workstations and printers? ☐ ☐ ☐
   d) OPAC workstations and printers? ☐ ☐ ☐
   e) Microform reader/printers? ☐ ☐ ☐
   f) Podiums? ☐ ☐ ☐
   g) Public address systems? ☐ ☐ ☐
   h) Tables? ☐ ☐ ☐
   i) Chairs? ☐ ☐ ☐
   j) Lounge furniture? ☐ ☐ ☐

k) Shelving for books and media? ☐ ☐ ☐
l) Video monitors? ☐ ☐ ☐
m) Video recorders ☐ ☐ ☐
n) Video players? ☐ ☐ ☐
o) Projection television? ☐ ☐ ☐
p) Screen (wall or rear view)? ☐ ☐ ☐
q) Equipment that may be obsolete, but might be required to play
media of interest to the library and/or its customers? Equipment might
include audiocassette players, VHS machines, and so on. ☐ ☐ ☐

Comments: _____

_____

9. Does the facility employ an in-the-floor grid system to accommodate and easily
change connections for electrical service, television, and communications distribution
throughout the multimedia area? ☐ ☐ ☐

Comments: _____

_____

10. Does the library have a gaming collection and room? This may include console
gaming stations available for individuals and groups wishing to play games in the
media library, PC game stations for the study of games and gaming, table space for
study, board games, and projects, and carrels with electrical outlets for private use.

Comments: _____

_____

11. Do the gaming stations have the same use policies as the computers in the library? ☐ ☐ ☐

Comments: _____

_____

F. Media Production and Presentation Labs

1. Does the library have a media production lab? Labs may provide computers, software,
and staff to assist users in the creation of multimedia-enhanced educational projects. ☐ ☐ ☐

Comments: _____

_____

2. Does the media center have a variety of equipment available for checkout to
certified users? ☐ ☐ ☐

Comments: _____

_____

3. In order to check out the equipment, do users have to demonstrate competencies
with the equipment and complete an equipment certification process in order to
check out equipment? ☐ ☐ ☐

Comments: _____

_____

	YES	NO	N/A

4. Does the lab staff provide training to library users? If so, does the facility have a training lab? ☐ ☐ ☐

   Comments: _____

   _____

5. Does the lab provide trained staff to assist customers with equipment, software, project planning, and feedback to ensure that all projects turn out as planned? ☐ ☐ ☐

   Comments: _____

   _____

6. Does the lab give library users access to high-end digital imaging workstations, slide and flatbed scanners, and a host of software to help digitize images and text for the web, archiving, or any other digital presentation format? ☐ ☐ ☐

   Comments: _____

   _____

7. Does the lab offer a variety of tools to help assemble documents for print or digital output? ☐ ☐ ☐

   Comments: _____

   _____

8. Does the media lab facilitate desktop publishing by offering a variety of tools to help assemble documents for print or digital output? ☐ ☐ ☐

   Comments: _____

   _____

G. Special Collections/Rare Books/Archives

1. Do the building program and/or institutional guidelines spell out the security necessary in the room? ☐ ☐ ☐

   Comments: _____

   _____

2. Is there a desk strategically located to allow an attendant a clear view of the readers? ☐ ☐ ☐

   Comments: _____

   _____

3. Is the reading room arranged to assure staff observance of those who are exiting the room? ☐ ☐ ☐

   Comments: _____

   _____

4. Are reading tables arranged in open positions, allowing maximum supervision from staff areas? ☐ ☐ ☐

   Comments: _____

   _____

5. Are the reading tables generously sized individual tables with task lighting, power
    for typewriters and/or laptop computers, and table lecterns for holding large books?   ☐   ☐   ☐

    Comments: _____

    _____

6. Are a few larger tables provided for use of large folios?   ☐   ☐   ☐

    Comments: _____

    _____

7. Are the rare books housed in locked cases with grilled doors?   ☐   ☐   ☐

    Comments: _____

    _____

8. Are the rare books shelved in specially designed (padded) book stacks that are securely
    braced with earthquake safety devices that prevent books from falling off shelves?   ☐   ☐   ☐

    Comments: _____

    _____

9. Are manuscripts and archives housed in acid-free boxes?   ☐   ☐   ☐

    Comments: _____

    _____

10. Are microfilm reading machines and other equipment provided to "read" all of the
    types of media and materials located in the room?   ☐   ☐   ☐

    Comments: _____

    _____

11. Are reading and exhibit areas separated?   ☐   ☐   ☐

    Comments: _____

    _____

12. Can an even temperature of 70 degrees F and humidity of about 50 percent be
    maintained to prolong the life of the books and materials?   ☐   ☐   ☐

    Comments: _____

    _____

13. Is there an electrostatic filter for the removal of dust and dirt?   ☐   ☐   ☐

    Comments: _____

    _____

14. In addition, is there a backup mechanical filter should the electrostatic filter break down?   ☐   ☐   ☐

    Comments: _____

    _____

15. Is the location of the air intake high enough on the exterior wall or roof to avoid
    chemical and exhaust pollution, especially for libraries sited in urban areas?   ☐   ☐   ☐

    Comments: _____

    _____

	YES	NO	N/A

16. Is care taken to control the levels of damaging (especially ultraviolet) light? ☐ ☐ ☐

Comments: _____

_____

17. Is the area monitored for insects, rodents, and other biological pests that may attack the collection? (Mechanical and/or chemical control techniques can be used.) ☐ ☐ ☐

Comments: _____

_____

18. Is the area monitored and protected to provide security with:

   a) A vault or strong room? ☐ ☐ ☐
   b) Special restricted keying and access? ☐ ☐ ☐
   c) Intrusion alarms? ☐ ☐ ☐
   d) Door contacts and other forms of perimeter protection? ☐ ☐ ☐
   e) Smoke and fire alarms? ☐ ☐ ☐
   f) Monitoring controls and alarms to indicate changes from desired temperature and/or humidity? ☐ ☐ ☐
   g) Water alarms? ☐ ☐ ☐
   h) Special alarms in display cases? ☐ ☐ ☐
   i) Panic alarms for staff? ☐ ☐ ☐
   j) Security video cameras to monitor the collection and reading areas? ☐ ☐ ☐

Comments: _____

_____

19. Has the library instituted a policy as to how to respond to alarms and where their signals should be seen or heard? ☐ ☐ ☐

Comments: _____

_____

20. Is there adequate workspace provided for conservation work? ☐ ☐ ☐

Comments: _____

_____

21. Is a disaster preparedness plan in effect? ☐ ☐ ☐

Comments: _____

_____

H. Reserve Book Room

1. Does the library have a reserve book room? The reserve book room provides access to course-related books which faculty has placed in the library for their students to borrow for short periods of time. ☐ ☐ ☐

Comments: _____

_____

	YES	NO	N/A

2. What is kept in the reserve book room?

   a) Items selected by instructors that are needed by students for class assignments? ☐ ☐ ☐

   b) High-demand items selected by the university library to allocate use? ☐ ☐ ☐

Comments: _____

_____

3. Can the materials in the reserved book room be checked out? ☐ ☐ ☐

   a) If yes, for how long may the materials be checked out? ☐ ☐ ☐

   b) Are there multiple loan periods? ☐ ☐ ☐

Comments: _____

_____

4. Are electronic reserves available on workstations in the room? An electronic reserve system is a web-based system for articles, course notes, and other supplemental materials. The library usually takes care of scanning, document management, and copyright permissions. ☐ ☐ ☐

Comments: _____

_____

5. Are the paper reserves kept in closed stacks accessible only by the staff? ☐ ☐ ☐

Comments: _____

_____

6. Are paper reserves accessible by the public on open shelves in the reserve book area? ☐ ☐ ☐

Comments: _____

_____

7. Is the reserve book room open for different periods than the main library? If so, how is access obtained to the rare book room when the main library is closed? ☐ ☐ ☐

Comments: _____

_____

8. Is there duplication equipment in the reserve book room for copying of documents and books? ☐ ☐ ☐

Comments: _____

_____

I. Faculty/Graduate Carrels and Study Rooms

1. Are there library carrels and small study rooms that may be reserved for extended periods of time by faculty and graduate students? ☐ ☐ ☐

Comments: _____

_____

	YES	NO	N/A

2. May library users have exclusive use of the carrel by reserving it for a semester or academic year? ☐ ☐ ☐

   Comments: _____

   _____

3. Has a means of allocating study rooms and carrels been established? ☐ ☐ ☐

   Comments: _____

   _____

4. How long may the study room and carrels be reserved, and how may they be renewed? ☐ ☐ ☐

   Comments: _____

   _____

5. Are there key locks on the assigned carrels and study rooms? ☐ ☐ ☐

   Comments: _____

   _____

6. If there are key locks, who controls the issuance of the keys? ☐ ☐ ☐

   Comments: _____

   _____

7. Do books used in the carrels and study rooms have to be checked out if they are kept in the room? ☐ ☐ ☐

   Comments: _____

   _____

J. Group, Quiet, and Silent Study Spaces

1. Does the library have a group study space? This is an area for students working in groups where distractions are expected. ☐ ☐ ☐

   a) Does the space require booking in advance of use? ☐ ☐ ☐
   b) Does the group study space allow collaborative study? ☐ ☐ ☐
   c) Are people in the space permitted to carry on regular conversations? ☐ ☐ ☐
   d) Are cell phones required to be on vibrate? ☐ ☐ ☐
   e) Are headphones required for viewing online videos or listening to electronic devices? ☐ ☐ ☐

   Comments: _____

   _____

2. Does the library have a quiet study space? This is an area with minimal distractions for students and/or library customers. ☐ ☐ ☐

   a) Is the quiet study space for individual study only? ☐ ☐ ☐
   b) Does it allow whispered or low-level conversations? ☐ ☐ ☐

	YES	NO	N/A

c)   Are cell phones required to be on vibrate?   ☐ ☐ ☐

d)   Are headphones required for viewing online videos or listening to electronic devices?   ☐ ☐ ☐

Comments: _____

_____

3.   Does the library have a silent study space? This is an area for students who wish to study alone with no distractions.

a)   Is the silent study space for individual study only?   ☐ ☐ ☐

b)   Does the space prohibit any conversation?   ☐ ☐ ☐

c)   Are cell phones required to be on vibrate?   ☐ ☐ ☐

d)   Does the area prohibit the viewing of online videos or use of electronic devices?   ☐ ☐ ☐

Comments: _____

_____

4.   Is there a system to report noise disturbances to a library staff member via chat, text message, or in person at a customer service desk?   ☐ ☐ ☐

Comments: _____

_____

K. Literacy Center

1.   Does the library provide literacy or reading center service?   ☐ ☐ ☐

Comments: _____

_____

2.   Is the literacy center a separate room or area in the library?   ☐ ☐ ☐

Comments: _____

_____

3.   Is there office space and equipment for the literacy program manager?   ☐ ☐ ☐

Comments: _____

_____

4.   Is space provided for a public bulletin display board and brochure rack?   ☐ ☐ ☐

Comments: _____

_____

5.   Is there space for a literacy book collection?   ☐ ☐ ☐

Comments: _____

_____

6.   Are there two-position tutoring study carrels for learner and tutor interaction?   ☐ ☐ ☐

Comments: _____

_____

	YES	NO	N/A

7. Is there space for computer learning stations? (Each station should have seating for two [learner and tutor], a computer with appropriate software, and storage for software and supplies.)   ☐ ☐ ☐

Comments: _____

_____

8. Is the lab acoustically controlled so that noise will not impact learners using the lab or other areas of the learning center?   ☐ ☐ ☐

Comments: _____

_____

9. Is there a small conference room that might serve as a functional office as well as a place for informal discussion?   ☐ ☐ ☐

Comments: _____

_____

10. Are there workstations or work areas for staff?   ☐ ☐ ☐

Comments: _____

_____

11. Are there workstations or work areas for volunteers?   ☐ ☐ ☐

Comments: _____

_____

L. Young Adult Facilities

1. Did a teen advisory panel work with the design team in developing the young adult space?   ☐ ☐ ☐

Comments: _____

_____

2. Is the location of the young adult area easily determined when one enters the library?   ☐ ☐ ☐

Comments: _____

_____

3. Is the young adult section separate from other areas in the library?   ☐ ☐ ☐

Comments: _____

_____

4. Is the space closer to the adult section than to the children's section?   ☐ ☐ ☐

Comments: _____

_____

5. Does the space encourage young adult use by allowing them to "control it" as they control personal space in their homes?   ☐ ☐ ☐

Comments: _____

_____

6. Is the space slightly secluded, giving the appearance of privacy, while still allowing some supervision?

☐ ☐ ☐

Comments: _____

_____

7. Does the space include glassed-in and acoustically separate seminar rooms that allow group study?

☐ ☐ ☐

Comments: _____

_____

8. Does the space include a glassed-in and acoustically separate area with a large-screen television and audio equipment?

☐ ☐ ☐

Comments: _____

_____

9. Do the materials housed in the young adult area appeal to the intended audience? Materials such as paperbacks in multiple copies arranged as in bookstores, uncluttered shelves, and collections grouped by genre such as science fiction, romances, and mysteries?

☐ ☐ ☐

Comments: _____

_____

10. Are the shelving and fixtures used to store young adult materials similar to those found in music, video, and bookstores?

☐ ☐ ☐

Comments: _____

_____

11. Does the space include computers for word processing and spreadsheets, access to the Internet, and games?

☐ ☐ ☐

Comments: _____

_____

12. Is there secure and adequate space to store teen gear such as skateboards and backpacks?

☐ ☐ ☐

Comments: _____

_____

13. Does the space allow a variety of comfortable seating options including traditional seating, chairs designed to tilt back without tipping, couches, and floor seating?

☐ ☐ ☐

Comments: _____

_____

14. Is there space allocated to reflect young adult pride and activities including bulletin boards listing teen accomplishments and activities?

☐ ☐ ☐

Comments: _____

_____

M. Children's Facilities

1. Is the physical and psychological environment pleasant and inviting to children? If you were a child, would this area appeal to you? ☐ ☐ ☐

   Comments: _____

   _____

2. Is there a policy in place for adult access to the children's facilities? ☐ ☐ ☐

   Comments: _____

   _____

3. Is the children's area arranged in such a manner that adults are not reluctant to use it? ☐ ☐ ☐

   Comments: _____

   _____

4. If there is a children's staff office, is it of adequate size? ☐ ☐ ☐

   Comments: _____

   _____

5. Is there a separate children's card catalog or an online public access terminal? ☐ ☐ ☐

   Comments: _____

   _____

6. Are shelving and furniture scaled for children? ☐ ☐ ☐

   Comments: _____

   _____

7. Is there furniture to accommodate plus-sized children? ☐ ☐ ☐

   Comments: _____

   _____

8. Are there small alcoves, surrounded by low shelves, controllable by the staff but accessible to children, where the children may pick out a book or game to settle individually or in small groups to enjoy it? ☐ ☐ ☐

   Comments: _____

   _____

9. Are the drinking fountains scaled for children? ☐ ☐ ☐

   Comments: _____

   _____

10. Are there restrooms scaled for children in the children's area? ☐ ☐ ☐

    Comments: _____

    _____

11. Do one or more of the children's restrooms include a diaper-changing table? ☐ ☐ ☐

    Comments: _____

    _____

12. If restroom facilities are not located in the children's area, are they located adjacent to or near the children's area? ☐ ☐ ☐

    Comments: _____

    _____

13. Are there some imaginative pieces of furniture for visual surprise? ☐ ☐ ☐

    Comments: _____

    _____

14. Are cheerful colors, interesting geometric shapes, and graphic sketches used in the children's area? ☐ ☐ ☐

    Comments: _____

    _____

15. Have sharp corners and edges been eliminated from furniture and equipment? ☐ ☐ ☐

    Comments: _____

    _____

16. Are the tabletops, chairs, and floors easily cleaned? ☐ ☐ ☐

    Comments: _____

    _____

17. Is there comfortable adult seating for use while adults are sharing books with children? ☐ ☐ ☐

    Comments: _____

    _____

18. Does the staff have visual control of the area? ☐ ☐ ☐

    Comments: _____

    _____

19. Are realia conveniently and attractively housed? ☐ ☐ ☐

    Comments: _____

    _____

20. Is there sufficient space for use and secure storage (locked if needed) of audiovisual materials and equipment? ☐ ☐ ☐

    Comments: _____

    _____

21. Is there sufficient space for crafts activities and storage of crafts materials? ☐ ☐ ☐

    Comments: _____

    _____

22. Is the floor a single height to allow for flexibility in programming and accessibility, as well as to avoid injuries? ☐ ☐ ☐

    Comments: _____

    _____

	YES	NO	N/A

23. Is there a separate programming area adjacent to, but out of, the traffic flow? ☐ ☐ ☐

   Comments: _____

   _____

24. Is the programming area designed to be multipurpose when not used for special functions, that is, quiet study, computer resource center, and so on? ☐ ☐ ☐

   Comments: _____

   _____

25. Is the programming area designed to handle the full age range of children who use the library? ☐ ☐ ☐

   Comments: _____

   _____

26. Has allowance been made for storage of special equipment used in programming, such as a puppet stage? ☐ ☐ ☐

   Comments: _____

   _____

27. Is the children's area acoustically designed to avoid interfering with other library functions? ☐ ☐ ☐

   Comments: _____

   _____

28. Do interior finishes and materials enhance the acoustics? ☐ ☐ ☐

   Comments: _____

   _____

29. Are play areas designed to avoid interfering with other library functions? ☐ ☐ ☐

   Comments: _____

   _____

30. Has allowance been provided for specific displays and materials geared to children? ☐ ☐ ☐

   Comments: _____

   _____

31. If children's and adult circulation counters are separated, is there lower counter space set aside for children, visibly marked by large graphics? ☐ ☐ ☐

   Comments: _____

   _____

32. Has sufficient space been allowed for easy access by children if materials are checked out or returned at the children's desk? ☐ ☐ ☐

   Comments: _____

   _____

N. Meeting and Seminar Rooms

1. Is the meeting room entry close to the main entrance?

    Comments: _____

    _____

2. Is there an assembly area adequate in size for handling the arrival and departure of large groups that may be attending meetings?

    Comments: _____

    _____

3. Can the meeting room area be closed off from the remainder of the library?

    Comments: _____

    _____

4. When the meeting room is closed off from the rest of the library, do users have access to public restrooms?

    Comments: _____

    _____

5. Are floor coverings easy to clean and replace?

    Comments: _____

    _____

6. Is a portable or built-in stage required?

    Comments: _____

    _____

7. Will a lectern or podium be required?

    Comments: _____

    _____

8. Is there a public telephone that may be used when the library is closed?

    Comments: _____

    _____

9. Is there a drinking fountain that may be used when the library is closed?

    Comments: _____

    _____

10. If the meeting room is large, is it equipped with folding doors that can be used as dividers to split the room into two or more parts?

    Comments: _____

    _____

	YES	NO	N/A

11. If folding partitions are used, can users get to and from each meeting room without disturbing those in adjacent rooms? ☐ ☐ ☐

   Comments: _____

   _____

12. Are there provisions for hanging coats and other personal gear? ☐ ☐ ☐

   Comments: _____

   _____

13. Does the room provide flexibility to accommodate a variety of programming activities, from children's story hours to film showings to art exhibitions? ☐ ☐ ☐

   Comments: _____

   _____

14. Are there special lighting fixtures and dimmer switches located in the ceiling above the speaker to provide glare-free and appropriate lighting? ☐ ☐ ☐

   Comments: _____

   _____

15. Is the lighting controllable in intensity, allowing full darkening of the room for visual presentations? ☐ ☐ ☐

   Comments: _____

   _____

16. Are window coverings provided to darken the room and block out light for visual presentations? ☐ ☐ ☐

   Comments: _____

   _____

17. Is there a kitchenette for the preparation of food and for serving light refreshments? ☐ ☐ ☐

   Comments: _____

   _____

18. Are there provisions for lockable pass-through from the kitchen to the meeting room for food and beverage service? ☐ ☐ ☐

   Comments: _____

   _____

19. Is the kitchen equipped with a sink, garbage disposal, microwave oven, stove, refrigerator, ice maker, and cabinets for storage of dishes and equipment? ☐ ☐ ☐

   Comments: _____

   _____

20. Is the room appropriately wired for phone, cable, teleconferencing, and so on? ☐ ☐ ☐

   Comments: _____

   _____

21. Is there a policy in place to discourage the use of cell phones in meeting areas? Control of cell phone nuisance may lie in promoting old-fashioned manners because cell phone jammers are prohibited in many localities.   ☐ ☐ ☐

Comments: _____

_____

22. Are there electrical and telecommunication outlets on all walls and at needed locations on the floor?   ☐ ☐ ☐

Comments: _____

_____

23. Are adequate space, data lines, and power provided for the following equipment:

   a) Overhead projectors?   ☐ ☐ ☐
   b) Projection from notebook and built-in computers?   ☐ ☐ ☐
   c) 35mm slide projectors?   ☐ ☐ ☐
   d) Ceiling- or wall-mounted screens?   ☐ ☐ ☐
   e) Large-screen or projection televisions?   ☐ ☐ ☐
   f) Video recorders and players?   ☐ ☐ ☐
   g) Teleconferencing equipment?   ☐ ☐ ☐
   h) Audio sound system using radio, MP3 players, and compact discs?   ☐ ☐ ☐
   i) Public address system?   ☐ ☐ ☐
   j) Wireless microphones?   ☐ ☐ ☐
   k) Podium with links to the various sound systems?   ☐ ☐ ☐
   l) Cable and/or satellite dish equipment?   ☐ ☐ ☐

Comments: _____

_____

24. Is there lockable storage for equipment?   ☐ ☐ ☐

Comments: _____

_____

25. Are there blackboards and/or white marker boards?   ☐ ☐ ☐

Comments: _____

_____

26. Are there electronic boards?   ☐ ☐ ☐

Comments: _____

_____

27. Are there art rails for exhibitions?   ☐ ☐ ☐

Comments: _____

_____

28. Are caddies available to move and store the chairs?   ☐ ☐ ☐

Comments: _____

_____

	YES	NO	N/A

29. Are the tables folding? □ □ □

Comments: _____

_____

30. Does the folding mechanism of the tables operate easily? □ □ □

Comments: _____

_____

31. Do the tables have:

a) Fixed-height bases? □ □ □
b) Adjustable-height bases? □ □ □

Comments: _____

_____

32. Are caddies available to store and move the tables? □ □ □

Comments: _____

_____

33. Are the chairs and tables light enough to be moved and maneuvered by library staff? □ □ □

Comments: _____

_____

34. Are there lockable storage areas near meeting rooms for audiovisual equipment and/or furniture such as lecterns or stackable chairs? □ □ □

Comments: _____

_____

35. Have provisions been made to prevent noisy programs from interfering with library operations? □ □ □

Comments: _____

_____

O. Convenience Facilities

1. Are restrooms located close to the lobby or building entrance? □ □ □

Comments: _____

_____

2. Does every floor have restrooms for both men and women? □ □ □

Comments: _____

_____

3. Are transgender restrooms available? □ □ □

Comments: _____

_____

4. Do single-occupancy restroom facilities use gender-neutral signage for those facilities? For example, by replacing signs that indicate "Men" and "Women" with signs that say "Restroom." ☐ ☐ ☐

   Comments: _____

   _____

5. Are transgender individuals allowed to use a gender-identity appropriate restroom without being harassed or questioned? ☐ ☐ ☐

   Comments: _____

   _____

6. Is a transitioning individual allowed to use the facility based on their current gender presentation? Specifically, their reassigned gender following commencement of the "real life experience" and from that point forward. ☐ ☐ ☐

   Comments: _____

   _____

7. Are restrooms easily identified? ☐ ☐ ☐

   Comments: _____

   _____

8. Are there special restrooms for children, located in or near the children's area? ☐ ☐ ☐

   Comments: _____

   _____

9. Do all restrooms contain an area for changing children's diapers? ☐ ☐ ☐

   Comments: _____

   _____

10. Does every floor have a drinking fountain? ☐ ☐ ☐

   Comments: _____

   _____

11. Are there drinking fountains for children? ☐ ☐ ☐

   Comments: _____

   _____

12. Are public telephones available? ☐ ☐ ☐

   Comments: _____

   _____

13. If public telephones are available, are they strategically located to encourage convenient use while preventing disturbance to other customers? ☐ ☐ ☐

   Comments: _____

   _____

	YES	NO	N/A

14. Is space allocated for public access to:

a) Photocopiers? ☐ ☐ ☐
b) Tele-facsimile (fax) machines? ☐ ☐ ☐
c) Personal computers? ☐ ☐ ☐
d) Computer printers? ☐ ☐ ☐
e) Audiovisual equipment? ☐ ☐ ☐
f) Other? ☐ ☐ ☐

Comments: _____

_____

15. Are signs available identifying these machines? ☐ ☐ ☐

Comments: _____

_____

16. Are coin-changing machines located near these machines? ☐ ☐ ☐

Comments: _____

_____

17. Does the library sell a "credit-type card" for paying for machine services? ☐ ☐ ☐

Comments: _____

_____

18. Are provisions made for noise abatement in noisy areas of the library? ☐ ☐ ☐

Comments: _____

_____

19. Are provisions made for trash and recycling? ☐ ☐ ☐

Comments: _____

_____

20. Is there a refreshment area available for the public? ☐ ☐ ☐

Comments: _____

_____

21. Are vending machines available for public use? ☐ ☐ ☐

Comments: _____

_____

22. Is the refreshment area located away from public service areas? ☐ ☐ ☐

Comments: _____

_____

23. Is the refreshment area easily viewed and supervised by staff? ☐ ☐ ☐

Comments: _____

_____

	YES	NO	N/A

24. Are trash receptacles available? ☐ ☐ ☐

    Comments: _____

    _____

25. Are clocks strategically located and visible in every major public area? ☐ ☐ ☐

    Comments: _____

    _____

26. Are the clocks controlled electronically to centrally adjust if needed? If not, are they easily accessible for resetting the time? ☐ ☐ ☐

    Comments: _____

    _____

27. If smoking is permitted, are smoking areas clearly identified? ☐ ☐ ☐

    Comments: _____

    _____

28. Is there a separate elevator for staff? ☐ ☐ ☐

    Comments: _____

    _____

29. Is there a separate elevator for freight? It is strongly recommended to have a large freight elevator in multistory buildings. ☐ ☐ ☐

    Comments: _____

    _____

P. Displays

1. Does the library provide smartphone apps to give visitors tools for an enhanced visiting experience? ☐ ☐ ☐

   Comments: _____

   _____

2. Do the apps incorporate text, audio, video, and other services like location systems? ☐ ☐ ☐

   Comments: _____

   _____

3. Does the library use quick response (QR) codes? These are a type of bar code that is readable with a QR code reader. It looks like a black-and-white square with lots of squiggly lines. ☐ ☐ ☐

   Comments: _____

   _____

4. Are the display furnishings and shelving appropriate for merchandising the library's products and services? ☐ ☐ ☐

   Comments: _____

   _____

	YES	NO	N/A

5. Can library materials be arranged in an attractive, appealing way to promote library products? ☐ ☐ ☐

Comments: _____

_____

6. Does the display shelving have built-in signs, boards, and lights to draw the attention of the library user? ☐ ☐ ☐

Comments: _____

_____

7. Are there racks for displaying audiovisual materials that are stable when filled? ☐ ☐ ☐

Comments: _____

_____

8. Are there bulletin boards for community notices and activities? ☐ ☐ ☐

Comments: _____

_____

9. Are there secure and locked exhibit cases, both free-standing and built-in? ☐ ☐ ☐

Comments: _____

_____

10. Do the cases have non-glare lighting to highlight the exhibits? ☐ ☐ ☐

Comments: _____

_____

11. Do the cases have surfaces that make posting easy? ☐ ☐ ☐

Comments: _____

_____

12. Are the cases ventilated to avoid overheating and damaging the exhibits? ☐ ☐ ☐

Comments: _____

_____

13. Are display cases located in high-traffic areas to make these areas more visually interesting? ☐ ☐ ☐

Comments: _____

_____

14. Is there space for the distribution of community information, tax forms, flyers, and other handouts? ☐ ☐ ☐

Comments: _____

_____

15. Is there a clear modular system of racks and displays for distribution of community and/or notices and giveaway items to prevent clutter? ☐ ☐ ☐

Comments: _____

_____

16. Are the racks and displays for distributing materials flexible enough to handle a variety of sizes and shapes of literature in a neat, attractive manner? ☐ ☐ ☐

   Comments: _____

   _____

17. Are the racks displayed in highly visible locations in order to attract customers and merchandise materials? ☐ ☐ ☐

   Comments: _____

   _____

Q. Public Art

1. Will public art be integrated into the new library or library expansion/renovation? Public art refers to works of *art* in any *media* that has been planned and executed with the specific intention of being sited or staged in or around the library. It is usually permanent or long-term art, and not a temporary exhibit. ☐ ☐ ☐

   Comments: _____

   _____

2. Does the library have a policy covering gifts of art that a donor wants to give to the library? ☐ ☐ ☐

   Comments: _____

   _____

3. Has the library allocated part of the construction project for public art? Usually this is a percentage of the total construction cost. ☐ ☐ ☐

   Comments: _____

   _____

4. If art is not part of the construction project, has funding been allocated for artwork from other sources? ☐ ☐ ☐

   Comments: _____

   _____

5. In a construction project, is selection of artists done by:

   a) Direct election where the artist is selected directly by the library or a committee from the university, political jurisdiction, board, and so on? ☐ ☐ ☐

   b) Limited competition where the artist is invited by the library, university, political jurisdiction, board, and so on to submit proposals? ☐ ☐ ☐

   c) Open competition where any artist or design team applies in response for a request for proposal/qualification and is selected by the library, university, political jurisdiction, board, and so on? ☐ ☐ ☐

   d) As part of the architects team as a sub-consultant? ☐ ☐ ☐

   Comments: _____

   _____

6. If the library issues a request for proposals (RFP) for architects:

   a) Is the RFP advertised in local newspapers, newsletters, direct mail, or other approved methods?   ☐ ☐ ☐

   b) Does the RFP indicate that public art is part of the construction project?   ☐ ☐ ☐

   c) Is there a committee in place to screen responses to the RFP?   ☐ ☐ ☐

   d) Who is responsible for the ultimate selection of the artist? _____

   Comments: _____

7. In making selection of the artist or artists, are the following factors considered in the selection process:

   a) Completeness of response to the RFP? Does the artist address each issue as set forth in the RFP?   ☐ ☐ ☐

   b) Quality? Is the highest priority given to the artist with the best-quality artwork?   ☐ ☐ ☐

   c) Media? Are all types of media considered for public art?   ☐ ☐ ☐

   d) Styles? Are all styles of art considered?   ☐ ☐ ☐

   e) Nature? Is the art appropriate in scale, material, form, and content for the physical space being considered for the art?   ☐ ☐ ☐

   f) Permanence? In addition to the quality of the art, will the art be relatively permanent, protected against theft and vandalism, easy to maintain, and will it stand up well to ambient conditions?   ☐ ☐ ☐

   g) Will the art add to the architectural project? Will the art enhance the architecture of the new library or the library expansion?   ☐ ☐ ☐

   h) Engineered correctly? Is the art engineered for safety and technical feasibility?   ☐ ☐ ☐

   i) Does the art meet the standards and policies of the library?   ☐ ☐ ☐

   Comments: _____

8. Does the artist's involvement in the construction project include:

   a) Being involved at the inception of the project so that the art is integral to the design?   ☐ ☐ ☐

   b) Are the artist's concepts included when the overall design of the project is first presented to the client and the respective community?   ☐ ☐ ☐

   c) Are the artist's ideas incorporated into the construction documents and bid as an integral part of the construction?   ☐ ☐ ☐

   d) Does the artist subcontract directly with the lead architect or engineer?   ☐ ☐ ☐

   e) Does the artist receive a fee for design and negotiate fees for oversight of fabrication and installation on an as-needed basis?   ☐ ☐ ☐

   Comments: _____

9. Is there a policy in place to govern the decommissioning and disposal of the art?   ☐ ☐ ☐

   Comments: _____

R. Interior Signage

1. Do signs meet ADA requirements (see section 7)?   ☐ ☐ ☐

    Comments: _____

    _____

2. Has the sign system been integrated into the building design and furniture selection process (architecture, color, etc.)?   ☐ ☐ ☐

    Comments: _____

    _____

3. Is there consistency in signage throughout the building? (Signs that serve the same function throughout the building should have the same shape, size, layout, type size, and placement.)   ☐ ☐ ☐

    Comments: _____

    _____

4. Are the signs of good design? (Typeface, size, spacing of letters, contrast, use of symbols, and color should all be considered.)   ☐ ☐ ☐

    Comments: _____

    _____

5. Are the sizes of signs proportional to distance from users and are signs sequentially positioned to facilitate self-service?   ☐ ☐ ☐

    Comments: _____

    _____

6. Are the signs well lighted, easy to read, and positioned for a clear view?   ☐ ☐ ☐

    Comments: _____

    _____

7. Do signs use terminology consistently? (Only one term should be applied to any one area, service, etc.)   ☐ ☐ ☐

    Comments: _____

    _____

8. Is the text of the sign clearly and accurately written in order to communicate the intended message effectively and positively?   ☐ ☐ ☐

    Comments: _____

    _____

9. Is the signage system flexible enough that, as conditions change, signs can be changed or moved easily?   ☐ ☐ ☐

    Comments: _____

    _____

	YES	NO	N/A

10. Is redundancy avoided? (Too many signs, all providing the same message, can be as bad as no sign at all.)   ☐ ☐ ☐

   Comments: _____

   _____

11. Are signs positioned and designed to avoid injuries (sharp corners, height, etc.)?   ☐ ☐ ☐

   Comments: _____

   _____

12. Are signs as vandal-proof as possible?   ☐ ☐ ☐

   Comments: _____

   _____

13. Is the exterior monument sign(s) identifying the library positioned so that it is easy to read when approaching the library? (A sign perpendicular to the road is easier to read than a sign parallel to the road.)   ☐ ☐ ☐

   Comments: _____

   _____

14. Is there a directory identifying major library services and their locations?   ☐ ☐ ☐

   Comments: _____

   _____

15. Are directional signs available leading customers to different departments and placed at logical decision points?   ☐ ☐ ☐

   Comments: _____

   _____

16. Are there signs on doors and at the entrances to departments to identify the function or service within that room or area?   ☐ ☐ ☐

   Comments: _____

   _____

17. Are there signs to highlight temporary collections and services or to announce events taking place in the library?   ☐ ☐ ☐

   Comments: _____

   _____

18. Are there signs that can be easily changed on the end panels of stacks to identify which books are shelved in that range?   ☐ ☐ ☐

   Comments: _____

   _____

19. Are there signs to provide critical information about regulations, warnings, procedures, instructions, and hours?   ☐ ☐ ☐

   Comments: _____

   _____

	YES	NO	N/A

20. Are instructional signs available for use of electronic workstations? ☐ ☐ ☐

Comments: _____

_____

S. Workroom/Offices

1. Are there individual workstations for all staff? ☐ ☐ ☐

Comments: _____

_____

2. Are there adequate workstations for library volunteers? ☐ ☐ ☐

Comments: _____

_____

3. Are workstations free from distractions? ☐ ☐ ☐

Comments: _____

_____

4. When required for team activities, are some workstations designed to foster communication among staff? ☐ ☐ ☐

Comments: _____

_____

5. Are there lockers and/or coat closets where personal items can be stored and secured for staff and volunteers? ☐ ☐ ☐

Comments: _____

_____

6. Is there adequate at-hand storage space? ☐ ☐ ☐

Comments: _____

_____

7. Is there a sick bay or a place where ill staff members may rest? ☐ ☐ ☐

Comments: _____

_____

8. Is there adequate space for equipment such as electronic workstations and television/DVD units? ☐ ☐ ☐

Comments: _____

_____

9. Is there adequate space for a variety of types of library storage? ☐ ☐ ☐

Comments: _____

_____

	YES	NO	N/A

10. Is there a locking storage unit or area to secure valuable equipment such as CD and DVD players, video projectors, cameras, and so on? ☐ ☐ ☐

Comments: _____

_____

11. Is there a locking storage unit to secure media and other expensive items during processing and prior to delivery to the public shelves? ☐ ☐ ☐

Comments: _____

_____

12. Is there adequate space for technical services operations? ☐ ☐ ☐

Comments: _____

_____

13. Are there adequate sorting shelves for the storage of returned library items? ☐ ☐ ☐

Comments: _____

_____

14. Is there an electronic workstation(s) with a printer(s) in the workroom to check in library items and look up records? ☐ ☐ ☐

Comments: _____

_____

15. Is there a typewriter in addition to the electronic workstation? ☐ ☐ ☐

Comments: _____

_____

16. Are there telephones? ☐ ☐ ☐

Comments: _____

_____

17. Are there enough electrical outlets for all required equipment? ☐ ☐ ☐

Comments: _____

_____

18. Do the rooms receive a robust wireless signal? ☐ ☐ ☐

Comments: _____

_____

19. Are there enough data lines? ☐ ☐ ☐

Comments: _____

_____

20. Is there adequate space for book trucks at workstations and for their storage when not in use? ☐ ☐ ☐

Comments: _____

_____

21. Is the workflow pattern and room arrangement effective and conducive to staff productivity?   ☐ ☐ ☐

    Comments: _____

    _____

22. Are environmental conditions such as lighting, HVAC, and acoustics adequate and comfortable?   ☐ ☐ ☐

    Comments: _____

    _____

23. Are managers' offices separate, in enclosed rooms, to ensure privacy?   ☐ ☐ ☐

    Comments: _____

    _____

24. Are there electronic workstations in the managers' offices for typing evaluations and other confidential types of materials?   ☐ ☐ ☐

    Comments: _____

    _____

25. Does the public have convenient access to the managers' offices for those managers who interact with the public?   ☐ ☐ ☐

    Comments: _____

    _____

26. Does the manager have convenient access to the workroom from his or her office?   ☐ ☐ ☐

    Comments: _____

    _____

27. Are there provisions for U.S. mail and newspaper delivery when the library is closed?   ☐ ☐ ☐

    Comments: _____

    _____

T. Staff Lounge

1. Are there lockers and/or coat closets where personal items can be stored and secured for staff and volunteers?   ☐ ☐ ☐

   Comments: _____

   _____

2. Is there a kitchen for the preparation of food and for serving light refreshments?   ☐ ☐ ☐

   Comments: _____

   _____

	YES	NO	N/A

3. Is the kitchen equipped with a sink, garbage disposal, microwave oven, stove, refrigerator, icemaker, coffee maker, and cabinets for storage of dishes and equipment? ☐ ☐ ☐

Comments: _____

_____

4. Is there provision for a ventilating system to eliminate strong food odors? ☐ ☐ ☐

Comments: _____

_____

5. Are there vending machines for food and soft drinks? ☐ ☐ ☐

Comments: _____

_____

6. Are there adequate numbers of tables and chairs to accommodate all staff that may be using the lounge at the same time? ☐ ☐ ☐

Comments: _____

_____

7. Is there a cot/sofa that can be used by the staff or customers in case of an emergency? ☐ ☐ ☐

Comments: _____

_____

8. Is there a window to look out on a quiet, pleasant scene? ☐ ☐ ☐

Comments: _____

_____

9. Does the staff have separate restrooms from the public? ☐ ☐ ☐

Comments: _____

_____

10. Is the staff lounge acoustically treated to eliminate the transfer of sound to and from adjacent public and staff areas? ☐ ☐ ☐

Comments: _____

_____

U. Friends of the Library

1. Does the library have a Friends of the Library group? ☐ ☐ ☐

Comments: _____

_____

2. Do the Friends of the Library accept donations such as books (used/new) and other items (puzzles, magazines, audiovisual)? ☐ ☐ ☐

   Comments: _____

   _____

3. Do the Friends of the Library have their own counter or worktable to sort donated items? ☐ ☐ ☐

   Comments: _____

   _____

4. Do the Friends of the Library have equipment such as carts and dollies available to handle large donations of books and media? ☐ ☐ ☐

   Comments: _____

   _____

5. Do the Friends of the Library have convenient storage for the above equipment and cardboard boxes for packing the donated items? ☐ ☐ ☐

   Comments: _____

   _____

6. Are the donations placed on shelves on the premises so the staff can view and select whatever they want to add to their collection, send to the branch book exchange, or sell? ☐ ☐ ☐

   Comments: _____

   _____

7. Is there storage space either on or off the premises for the donated items? ☐ ☐ ☐

   Comments: _____

   _____

8. Do the Friends of the Library operate a retail store? ☐ ☐ ☐

   Comments: _____

   _____

9. Do the Friends have an:

   a) Annual book sale? ☐ ☐ ☐
   b) Ongoing daily book sale? ☐ ☐ ☐

   Comments: _____

   _____

10. If the Friends have an ongoing daily sale, do they sell their items from:

    a) A store? ☐ ☐ ☐
    b) Some shelves in the library? ☐ ☐ ☐
    c) A book cart? ☐ ☐ ☐
    d) An area adjacent to their room? ☐ ☐ ☐

    Comments: _____

    _____

	YES	NO	N/A

11. Is the book sale area clearly marked by signs? ☐ ☐ ☐

Comments: _____

_____

12. Is the cash from the Friends' sales kept in a separate place so as not to get confused with the daily cash from library cash? ☐ ☐ ☐

Comments: _____

_____

13. Is there space in a prominent area allotted to the Friends for their newsletter and membership applications? ☐ ☐ ☐

Comments: _____

_____

14. Do the Friends have a mail slot to receive their membership applications, dues, donations, and so on? ☐ ☐ ☐

Comments: _____

_____

15. Do the Friends have a bulletin board for messages? ☐ ☐ ☐

Comments: _____

_____

16. Do the Friends have some space in the staff lounge to hang their coats and lockers or some other safe place to store their valuables? ☐ ☐ ☐

Comments: _____

_____

V. Library Store

1. If a library store is provided, is the store in a prominent location to attract the attention of customers as they walk by? ☐ ☐ ☐

Comments: _____

_____

2. If there is a store, will it be operated by the library or outsourced to a private vendor? ☐ ☐ ☐

Comments: _____

_____

3. Are there adequate signage, window displays, and other visual cues to market the library to potential customers? ☐ ☐ ☐

Comments: _____

_____

4. Are the circulation paths simple and logical? (Customers should be able to concentrate on the merchandise and not be worried about bumping into things.) ☐ ☐ ☐

Comments: _____

_____

5. Is the cash/wrapping counter designed and located for maximum efficiency, accessibility, and optimal equipment placement? ☐ ☐ ☐

    Comments: _____

    _____

6. Is there a cash register? ☐ ☐ ☐

    Comments: _____

    _____

7. Is there an office/storage room located in the store? ☐ ☐ ☐

    Comments: _____

    _____

8. Is the design of the display fixtures flexible to permit new products to be added periodically? ☐ ☐ ☐

    Comments: _____

    _____

9. Will some products require special displays or display techniques? ☐ ☐ ☐

    Comments: _____

    _____

10. Is a specific lighting source (incandescent, fluorescent, or halogen) preferred? ☐ ☐ ☐

    Comments: _____

    _____

11. Are there security systems in place to protect staff, merchandise, and cash? ☐ ☐ ☐

    Comments: _____

    _____

12. Are there special requirements for cooling or heating any areas of the store? ☐ ☐ ☐

    Comments: _____

    _____

13. Are telephone and data connections required? ☐ ☐ ☐

    Comments: _____

    _____

14. Are there enough electrical outlets? ☐ ☐ ☐

    Comments: _____

    _____

15. Will the store have a sound system? ☐ ☐ ☐

    Comments: _____

    _____

	YES	NO	N/A

16. Are there preferred materials for the walls? ☐ ☐ ☐

Comments: _____

_____

17. Are there preferred materials for the floors? ☐ ☐ ☐

Comments: _____

_____

18. Are there preferred materials for the ceiling? ☐ ☐ ☐

Comments: _____

_____

W. Interior Storage

1. Has storage been considered in planning the library? ☐ ☐ ☐

Comments: _____

_____

2. Is there a room to store pieces of furniture, equipment, displays, and other miscellaneous items? ☐ ☐ ☐

Comments: _____

_____

3. Is there storage space for less frequently used library materials such as old newspapers, periodicals, and donated books awaiting review for possible addition to the collections? ☐ ☐ ☐

Comments: _____

_____

4. Is there adequate storage for office and library supplies? ☐ ☐ ☐

Comments: _____

_____

5. Is there another building or library property where infrequently used materials can be stored to make room for rapidly growing collections? ☐ ☐ ☐

Comments: _____

_____

6. Is there a policy in place to keep the storage area from becoming the library's "attic"? ☐ ☐ ☐

Comments: _____

_____

7

Compliance with ADA Accessibility Guidelines

A. Guidelines

1. The guidelines below are not comprehensive, and it is recommended that the Americans with Disabilities Act's (ADA's) "Standards for Accessible Design" and/or the U.S. Access Board, which are cited below, be consulted. The site for the ADA is www.ada.gov.

    Comments: _____

2. The best source for the latest information about ADA Requirements is the U.S. Department of Justice site, ADA Standards for Accessible Design: www.usdoj.gov/crt/ada/adastd94.pdf. This site also provides figures that illustrate many of the requirements below.

    Comments: _____

3. Another good site is provided by the U.S. Access Board, which is a federal agency that promotes equality for people with disabilities through leadership in accessible design and the development of accessibility guidelines and standards for the built environment: www.access-board.gov.

    Comments: _____

4. All areas of newly designed and newly constructed buildings and facilities and altered portions of existing buildings and facilities shall comply with ADA requirements.

    Comments: _____

B. Transportation, Parking Lots, Parking Signage, and Accessible Routes

1. Is there at least one accessible route within the boundary of the site from a public transportation stop? An accessible route is a continuous unobstructed path connecting all accessible elements and spaces of a building. Interior accessible routes may include corridors, floors, ramps, elevators, lifts, and clear floor space at fixtures. Exterior accessible routes may include parking access aisles, curb ramps, crosswalks at vehicular ways, walks, ramps, and lifts. ☐ ☐ ☐

   Comments: _____

   _____

2. Are there safe and accessible parking spaces located on the shortest accessible route of travel to an accessible entrance? ☐ ☐ ☐

   Comments: _____

   _____

3. If the building has multiple accessible entrances with adjacent parking, are accessible parking spaces dispersed and located closest to the accessible entrances? ☐ ☐ ☐

   Comments: _____

   _____

4. Do the accessible routes have floors walks, ramps, stairs, and curb ramps that are stable, firm and slip-resistant? ☐ ☐ ☐

   Comments: _____

   _____

5. Do accessible parking spaces have a designated sign showing the symbol of accessibility? ☐ ☐ ☐

   Comments: _____

   _____

6. Is the minimum clear width of an accessible route at least thirty-six inches? ☐ ☐ ☐

   Comments: _____

   _____

7. Are the accessible parking spaces at least 8 feet wide and 20 feet long? ☐ ☐ ☐

   Comments: _____

   _____

8. Are the accessible parking lot spaces and aisles level so that wheelchairs will not roll if left unattended while transferring persons to their vehicle? ☐ ☐ ☐

   Comments: _____

   _____

9. Is 1 in every 8 accessible parking spaces, but not less than 1 overall, served by an access aisle 96 inches in width, with signage that indicates "Van Accessible" under the accessibility symbol? ☐ ☐ ☐

   Comments: _____

   _____

	YES	NO	N/A

10. Does the facility observe the following requirements for parking spaces?   ☐ ☐ ☐

No. of Spaces	Minimum Accessible Spaces
1 to 25	1
26 to 50	2
51 to 75	3
76 to 100	4
101 to 150	5
151 to 200	6
201 to 300	7
301 to 400	8
401 to 500	9
501 to 1,000	2 percent of total
1,001 and over	20, plus 1 for each 100 over 1,000

Comments: _____

_____

11. Are access aisles between van parking spaces five feet in width, striped, and part of
an accessible route? (Two accessible parking spaces can share a common access aisle.)   ☐ ☐ ☐

Comments: _____

_____

12. Does the van-accessible parking space clear vertically to at least 9 feet 6 inches high?   ☐ ☐ ☐

Comments: _____

_____

13. If the library has a passenger-loading zone, does the zone have an access aisle 5 feet
wide and 20 feet long, adjacent and parallel to vehicle pull-up space?   ☐ ☐ ☐

Comments: _____

_____

C. Ground and Floor Surfaces

1. Are slip-resistant floors used throughout the building and are floor surfaces stable
and firm?   ☐ ☐ ☐

Comments: _____

_____

2. Are changes in the vertical level less than ¼ inch? If so, no edge treatment is required.   ☐ ☐ ☐

Comments: _____

_____

3. Are changes in floor level between ⅓ and ½ inch? If so, the change in the level must
be beveled with a slope no greater than 1:2.   ☐ ☐ ☐

Comments: _____

_____

	YES	NO	N/A

4. Are changes in level greater than ½ inch? If so, they require a ramp. ☐ ☐ ☐

   Comments: _____

   _____

5. If carpet or carpet tile is used, is it securely attached to the floor, and does it have a firm cushion pad, a level loop, and a maximum pile thickness of less than ½ inch? ☐ ☐ ☐

   Comments: _____

   _____

6. Do spaces of gratings located in walking surfaces have spaces no greater than ½ inch in one direction? ☐ ☐ ☐

   Comments: _____

   _____

7. Do gratings have elongated openings? If so, they must be placed so that the long dimension is perpendicular to the dominant direction of travel. ☐ ☐ ☐

   Comments: _____

   _____

8. Can a visually disabled individual who is using a cane detect objects protruding from the wall or floor? ☐ ☐ ☐

   Comments: _____

   _____

9. Is there clear and distinct contrast between the floors and walls to assist the visually impaired? ☐ ☐ ☐

   Comments: _____

   _____

D. Curb Cuts

1. Are there curb cuts or curb ramps at all curbs and walks on accessible routes to accessible entrances? ☐ ☐ ☐

   Comments: _____

   _____

2. Are there any curbs between the access aisle and the vehicle pull-up space? If so, are there cuts or curb ramps? ☐ ☐ ☐

   Comments: _____

   _____

3. If there are curb ramps, are they built so they do not extend into vehicle traffic lanes? ☐ ☐ ☐

   Comments: _____

   _____

	YES	NO	N/A

4. Is the least possible slope used for every ramp? ☐ ☐ ☐

   Comments: _____

   _____

5. If a curb ramp is located where pedestrians must walk across the ramp, is it
   protected by handrails or guardrails? ☐ ☐ ☐

   Comments: _____

   _____

6. If the curb ramp does not have guardrails, does it have flared sides? ☐ ☐ ☐

   Comments: _____

   _____

7. Does the curb ramp have a detectable warning extending the full width and depth
   of the ramp? ☐ ☐ ☐

   Comments: _____

   _____

8. Do curb cuts or curb ramps have a slope of 1:12 or less and flared sides with a slope
   of 1:10, which is the minimum requirement? ☐ ☐ ☐

   Comments: _____

   _____

9. If the pavement is not level, is the slope no more than 2 percent in all directions? ☐ ☐ ☐

   Comments: _____

   _____

10. Do the curb cuts or curb ramps provide drainage so that water will not be trapped
    after a storm? ☐ ☐ ☐

    Comments: _____

    _____

11. Are curb cuts or ramps thirty-six inches wide excluding the flared sides? ☐ ☐ ☐

    Comments: _____

    _____

12. Are curb ramps at marked crossings wholly contained within the markings, excluding
    any flared sides? ☐ ☐ ☐

    Comments: _____

    _____

13. If diagonal (or corner-type) curb ramps have returned curbs or other well-defined
    edges, are they parallel to the direction of pedestrian flow? ☐ ☐ ☐

    Comments: _____

    _____

14. Are any raised islands in crossing cut through level with the street or do they have curb ramps at both sides and a level area at least thirty-eight inches long between the curb ramps in the part of the island intersected by the crossings? ☐ ☐ ☐

   Comments: _____

   _____

E. Ramps

1. Is the least possible slope used for any ramp? Any part of an accessible route with a slope greater than 1:20 shall be considered a ramp. ☐ ☐ ☐

   Comments: _____

   _____

2. Is the maximum rise for any ramp less than thirty inches? ☐ ☐ ☐

   Comments: _____

   _____

3. Does the ramp have a minimum clear width of thirty-six inches? ☐ ☐ ☐

   Comments: _____

   _____

4. Do the ramp landings have:
   a) Level landings at the bottom and top of each ramp and each ramp run? ☐ ☐ ☐
   b) A landing at least as wide as the ramp run leading to it? ☐ ☐ ☐
   c) A landing length of a minimum of sixty inches clear? ☐ ☐ ☐
   d) Does the ramp change direction at landings? If so, the minimum landing size shall be 60 inches by 60 inches. ☐ ☐ ☐
   e) Have level landings at the top and bottom, at least as wide as the ramp? ☐ ☐ ☐
   f) Have landings at least sixty inches in length? ☐ ☐ ☐
   g) Is there a doorway located at a landing? If so, the area in front of the doorway (not automatic or power-assisted) requires a minimum maneuvering clearance at the door of forty-eight inches. ☐ ☐ ☐

   Comments: _____

   _____

5. Does the ramp run have a rise greater than 6 inches or a horizontal projection greater than 72 inches? If so, then it requires handrails on both sides. ☐ ☐ ☐

   Comments: _____

   _____

6. Are the cross slopes of ramp surfaces no greater than 1:50? ☐ ☐ ☐

   Comments: _____

   _____

	YES	NO	N/A

7. Are outdoor ramps and their approaches designed so that water will not accumulate on walking surfaces? ☐ ☐ ☐

Comments: _____

_____

F. Stairs

1. On any given flight of stairs, do all steps have uniform riser heights and uniform tread levels? ☐ ☐ ☐

Comments: _____

_____

2. Are the stair treads no less than eleven inches measured from riser to riser? (Open risers are not permitted.) ☐ ☐ ☐

Comments: _____

_____

3. Is the underside of nosings (the usually rounded edge of a stair) not abrupt? ☐ ☐ ☐

Comments: _____

_____

4. Is the radius of the curvature at the leading edge of the tread no greater than ½ inch? ☐ ☐ ☐

Comments: _____

_____

5. Are the risers sloped or do the undersides of the nosing have an angle not less than sixty degrees from the horizontal? ☐ ☐ ☐

Comments: _____

_____

6. Do the stairways have handrails at both sides of the stairs? If so they should comply with F5 above. ☐ ☐ ☐

Comments: _____

_____

G. Lifts and Elevators

1. If the building has more than one floor, do wheelchair users have access to an elevator? ☐ ☐ ☐

Comments: _____

_____

2. Are accessible elevators on an accessible route? ☐ ☐ ☐

Comments: _____

_____

	YES	NO	N/A

3. Is the elevator operation automatic? ☐ ☐ ☐

   Comments: _____

   _____

4. Do the elevators have an automatic self-leveling feature that will automatically bring the car to floor landings within a tolerance of ½ inch under rated loading to zero loading conditions? This self-leveling feature shall be automatic and independent of the operating devices and shall correct the over- or under-travel. ☐ ☐ ☐

   Comments: _____

   _____

5. Are call buttons in elevator lobbies and hallways centered at forty-two inches above the floor? ☐ ☐ ☐

   Comments: _____

   _____

6. Do the call buttons have a minimum of ¾ inch in the smallest dimension, and is the button designating the up direction placed on top? ☐ ☐ ☐

   Comments: _____

   _____

7. Do the call buttons have visual signals to indicate when each call is registered and when each call is answered? ☐ ☐ ☐

   Comments: _____

   _____

8. Is there a visible and audible signal provided at each hoist way entrance to indicate which car is answering a call? Audible signals shall sound once for the up direction and twice for the down direction, or have verbal annunciators that say "up" or "down." ☐ ☐ ☐

   Comments: _____

   _____

9. Are hall lantern fixtures mounted so that their centerline is at least seventy-two inches above the lobby floor? ☐ ☐ ☐

   Comments: _____

   _____

10. Are the signals visible from the vicinity of the hall call button? ☐ ☐ ☐

    Comments: _____

    _____

11. Do all elevator hoist way entrances have raised and Braille floor designations provided on both jams? The centerline of the characters shall be 60 inches above the finish floor, and are the characters at least 2 inches in height? ☐ ☐ ☐

    Comments: _____

    _____

12. Do elevator doors open and close automatically, and are they provided with a reopening device that will stop automatically if the door becomes obstructed by an object or person? ☐ ☐ ☐

    Comments: _____

    _____

13. Do elevator doors open fully in response to a car call for a minimum of three seconds? ☐ ☐ ☐

    Comments: _____

    _____

14. Does the floor area of elevator cars provide space for a wheelchair user to enter the car, maneuver within reach of controls, and exit from the car? ☐ ☐ ☐

    Comments: _____

    _____

15. Is the level of illumination at the car controls, platform, and car threshold and landing sill at least five footcandles? ☐ ☐ ☐

    Comments: _____

    _____

16. Do elevator car controls have the following features?

    a) Are all control buttons at least ¾ inch in their smallest dimension, and are either raised or flush? ☐ ☐ ☐

    b) Are all control buttons designated by Braille and raised standard alphabet characters for letters, Arabic characters for numerals, or standard symbols? ☐ ☐ ☐

    c) Are all floor buttons no higher than 54 inches above the finish floor for side approach and 48 inches for front approach? ☐ ☐ ☐

    d) Are emergency controls grouped at the bottom of the panel with their center-lines no less than thirty-five inches above the finish floor? ☐ ☐ ☐

    e) Are the control buttons for the emergency stop, alarm, door open, door close, main entry floor, and phone identified with tactile symbols? ☐ ☐ ☐

    Comments: _____

    _____

17. Is a visual car position indicator provided above the car control panel or over the door to show the position of the elevator in the hoist way? As the car passes or stops at a floor served by the elevators, the corresponding numerals shall illuminate, and an audible signal shall sound. ☐ ☐ ☐

    Comments: _____

    _____

18. Is there an emergency two-way communication system between the elevator and a point outside the hoist way? ☐ ☐ ☐

    Comments: _____

    _____

H. Doors

1. Do doorways have a minimum clear opening of 32 inches with the door open 90 degrees? Clear openings of doorways with swinging doors shall be measured between the face of the door and the stop, with the door open 90 degrees. ☐ ☐ ☐

   Comments: _____
   _____

2. Are thresholds raised less than ½ inches from the floor? ☐ ☐ ☐

   Comments: _____
   _____

3. Does the door allow maneuvering clearance for a person in a wheelchair? Maneuvering room will vary based on type of door. ☐ ☐ ☐

   Comments: _____
   _____

4. Do handles, pulls, latches, locks, and other operating devices on accessible doors have a shape that is easy to grasp with one hand and does not require tight grasping, tight pinching, or twisting of the wrist to operate? ☐ ☐ ☐

   Comments: _____
   _____

5. Are there easy-to-grip door handles using push-type, lever-operated, or U-type handles? ☐ ☐ ☐

   Comments: _____
   _____

6. Is hardware required for accessible doors mounted no higher than forty-eight inches above the finished floor? ☐ ☐ ☐

   Comments: _____
   _____

7. Do accessible doors allow delay-closing action of at least 3 seconds to move from an open position of 70 degrees to a point 3 inches from the latch measured to the leading edge of the door? ☐ ☐ ☐

   Comments: _____
   _____

8. Does the maximum force for pushing or pulling open a door meet the following standards?

   a) Interior hinged doors; 51 pounds of force (lbf). ☐ ☐ ☐
   b) Sliding or folding doors; 5 lbf. ☐ ☐ ☐

   Comments: _____
   _____

I. Entrances

1. Is a book drop available near the entrance? Is so, the design of the handle should be at 54 inches above the floor for wheelchair approaches from the side. To allow a forward approach to the book drop, the maximum is 48 inches above to floor to allow reaching from the wheelchair.   ☐ ☐ ☐

    Comments: _____

    _____

2. Are all building entrances accessible to the disabled? Revolving doors or turnstiles shall not be the only means of passage at an accessible entrance or along an accessible route.   ☐ ☐ ☐

    Comments: _____

    _____

3. Do all accessible entrance doors display a sign or sticker with the symbol for accessibility?   ☐ ☐ ☐

    Comments: _____

    _____

4. If some entrances are not accessible, are signs displayed directing people to accessible entrances?   ☐ ☐ ☐

    Comments: _____

    _____

5. Do library security gates have a clear minimum opening of thirty-two inches?   ☐ ☐ ☐

    Comments: _____

    _____

6. Is there an accessible gate or door provided adjacent to the turnstile or revolving door and is it designed to facilitate the same use pattern?   ☐ ☐ ☐

    Comments: _____

    _____

7. Do doorways have a minimum clear opening of 32 inches with the door open 90 degrees?   ☐ ☐ ☐

    Comments: _____

    _____

8. Is the minimum space between two hinged or pivoted doors in a series at least forty-eight inches plus the width of any door swinging into the space?   ☐ ☐ ☐

    Comments: _____

    _____

9. Do door thresholds not exceed ¾ inches in height for exterior sliding doors, or ½ inches for other types of doors?   ☐ ☐ ☐

    Comments: _____

    _____

10. Do handles, pulls, latches, locks, and other operating devices on accessible doors have a shape that is easy to grasp with one hand and does not require tight grasping, tight pinching, or twisting of the wrist to operate? ☐ ☐ ☐

    a) Lever-operated mechanisms, push-type mechanisms, and U-shaped handles are acceptable designs. ☐ ☐ ☐

    b) Hardware required for accessible door opening shall be mounted no higher than forty-eight inches. ☐ ☐ ☐

    Comments: _____

_____

11. If a door has a closer, is it adjusted so that from an open position of 70 degrees, the door will take at least 3 seconds to move to a point of 3 inches from the latch, measured to the leading edge of the door? ☐ ☐ ☐

    Comments: _____

_____

12. Is the maximum force for pushing or pulling open a door as follows:

    a) Fire door shall have a minimum opening force allowable by the appropriate administrative authority. ☐ ☐ ☐

    b) Interior hinged doors and sliding or folding doors must have a maximum force of 5 lbf. ☐ ☐ ☐

    Comments: _____

_____

J. Accessible Routes within the Building

1. Is there at least one accessible route connecting accessible buildings, facilities, elements, and spaces? An accessible route is a continuous unobstructed path connecting all accessible elements and spaces of a building. Interior accessible routes may include corridors, floors, ramps, elevators, lifts, and clear floor space at fixtures. ☐ ☐ ☐

    Comments: _____

_____

2. Is the minimum clear width of an accessible route 36 inches except at doorways? If a person in a wheelchair must make a turn around an obstruction, does the minimum clear width allow this to take place? Usually this is 60 inches by 60 inches. ☐ ☐ ☐

    Comments: _____

_____

3. If the accessible route is less than 5 feet in width, are there passing spaces of 5 feet at intervals of not more than 200 feet? ☐ ☐ ☐

    Comments: _____

_____

	YES	NO	N/A

4. Is there headroom clearance of at least eighty inches above the floor along accessible routes? ☐ ☐ ☐

Comments: _____

_____

5. Are there protruding objects (e.g., telephones, drinking fountains, and/or furniture) along these travel routes? ☐ ☐ ☐

Comments: _____

_____

6. If there are any protruding objects with their lead edges at or below 27 inches, do they leave a clear minimum path of 36 inches? ☐ ☐ ☐

Comments: _____

_____

7. Are ground and floor surfaces stable, firm, and slip-resistant? ☐ ☐ ☐

Comments: _____

_____

8. Does any accessible route serving any accessible space or element serve as a means of egress for emergencies or connect to an accessible area of rescue assistance? ☐ ☐ ☐

Comments: _____

_____

K. Drinking Fountains

1. Are the spouts on all drinking fountains no higher than thirty-six inches measured from the floor to the spout outlet? ☐ ☐ ☐

Comments: _____

_____

2. Are the spouts of drinking fountains and water coolers at the front of the unit and do they direct the flow in trajectory that is parallel or nearly parallel to the front of the unit? ☐ ☐ ☐

Comments: _____

_____

3. Does the water flow at least four inches high above the spout so a cup or glass can be placed under the water flow? ☐ ☐ ☐

Comments: _____

_____

4. If the drinking fountain is round or an oval bowl, is the spout positioned so the edge of the water is within three inches of the front of the fountain? ☐ ☐ ☐

Comments: _____

_____

	YES	NO	N/A

5. Are controls front-mounted, or if side-mounted, near the front edge? ☐ ☐ ☐

   Comments: _____

   _____

6. Are controls and operating mechanisms operable with one hand, and do not require tight grasping, pinching, or twisting of the wrist? ☐ ☐ ☐

   Comments: _____

   _____

7. If the accessible water fountain is wall- or post-mounted and has knee space, is the space at least 27 inches high, 30 inches wide, and 17 to 19 inches deep? ☐ ☐ ☐

   Comments: _____

   _____

8. Does the water fountain have a minimum clear floor space of 30 inches by 48 inches to allow a person in a wheelchair to approach the unit facing forward? ☐ ☐ ☐

   Comments: _____

   _____

9. Do freestanding or built-in drinking fountains with a clear space under them have a clear floor space of at least 30 inches by 48 inches that allows a person in a wheelchair to make parallel approaches to the unit? ☐ ☐ ☐

   Comments: _____

   _____

L. Toilet Rooms

1. Are there signs indicating the nearest accessible toilet room available for persons with disabilities? ☐ ☐ ☐

   Comments: _____

   _____

2. Is there at least one toilet room accessible for persons with disabilities? ☐ ☐ ☐

   Comments: _____

   _____

3. Is the toilet room marked with Braille signage? ☐ ☐ ☐

   Comments: _____

   _____

4. Is the door into the toilet room easily opened by the disabled? See the section on doors for requirements. ☐ ☐ ☐

   Comments: _____

   _____

M. Toilet Stalls

1. Is at least one stall accessible to a wheelchair, and does it display the international symbol of accessibility?   ☐ ☐ ☐

   Comments: _____

   _____

2. Is the accessible toilet stall on an accessible route?   ☐ ☐ ☐

   Comments: _____

   _____

3. Are there 60 inches of clear floor space in the restroom for a wheelchair to make a 180-degree turn?   ☐ ☐ ☐

   Comments: _____

   _____

4. Does the immediate area allow forty-eight inches of clear space to approach the stall door?   ☐ ☐ ☐

   Comments: _____

   _____

5. Is there a minimum clearance of thirty-six inches between all fixtures to an accessible stall?   ☐ ☐ ☐

   Comments: _____

   _____

6. Do toilets in accessible stalls meet the requirements listed above for them?   ☐ ☐ ☐

   Comments: _____

   _____

7. Are stall doors at least thirty-six inches wide?   ☐ ☐ ☐

   Comments: _____

   _____

8. Does the door of the stall open out?   ☐ ☐ ☐

   Comments: _____

   _____

9. Do standards toilet stalls have a minimum depth of 56 inches for wall-mounted toilets, and 59 inches for a floor-mounted toilet?   ☐ ☐ ☐

   Comments: _____

   _____

10. Is there a toe clearance of at least 9 inches above the floor for standard stalls? If the depth of the stall is greater than 60 inches, toe clearance is not required.   ☐ ☐ ☐

   Comments: _____

   _____

11. Is the toilet-paper holder located within easy reach from the toilet and at least 19 inches from the floor, with continuous paper flow? ☐ ☐ ☐

    Comments: _____

    _____

12. Are coat and purse hooks at a height of approximately forty-eight inches to make them convenient to wheelchair users? ☐ ☐ ☐

    Comments: _____

    _____

13. Are there grab bars located in the stall, and do they meet the requirements listed in the section on "Handrails and Grab Bars" below? ☐ ☐ ☐

    Comments: _____

    _____

N. Toilets

1. Does clear space for toilets not located in stalls centered on the wall with direct access have a 48-inch minimum entrance, a 66-inch clear floor space in front of the toilet, and a 36-inch minimum space one the wall behind the water closet? ☐ ☐ ☐

   Comments: _____

   _____

2. Does clear space for toilets not located in stalls and located on one side of the rear wall with direct access have a 60-inch minimum entrance, a 56-inch clear floor space in front of the toilet, and a 60-inch minimum space one the wall behind the toilet? ☐ ☐ ☐

   Comments: _____

   _____

3. Are there grab bars for the toilets not located in stalls, 33 to 36 inches on center above the floor with a minimum of 42 inches in length? ☐ ☐ ☐

   Comments: _____

   _____

4. Is the height of the toilet between 17 to 19 inches measured at the top of the toilet seat? Seats shall not be sprung to return to a lifted position. ☐ ☐ ☐

   Comments: _____

   _____

5. Are flush controls operated by hand or automatically? Is the control for flush valves mounted on the wide side of toilet areas no more than forty-four inches above the floor? ☐ ☐ ☐

   Comments: _____

   _____

	YES	NO	N/A

6. Are toilet paper dispensers installed within reach and located nineteen inches above the floor? ☐ ☐ ☐

   Comments: _____

   _____

7. Are other fixtures in the restroom (soap dispensers, towels, auto-dryers, sanitary- napkin dispensers, waste paper receptacles, etc.) located so the controls or dispensers are at a maximum of forty-eight inches from the floor? ☐ ☐ ☐

   Comments: _____

   _____

8. Are coat and purse hooks at a height of approximately forty-eight inches to make them convenient to wheelchair users? ☐ ☐ ☐

   Comments: _____

   _____

O. Urinals

1. Is the rim of the accessible urinal no more than seventeen inches above the floor? ☐ ☐ ☐

   Comments: _____

   _____

2. Is there a clear space of at least 30 inches by 48 inches in front of the urinals? ☐ ☐ ☐

   Comments: _____

   _____

3. Are flush controls either hand-operated or automatic, and are they mounted no more than forty-four inches off the floor? ☐ ☐ ☐

   Comments: _____

   _____

4. Are the controls and operating mechanisms capable of being operated with one hand and not require tight grasping, pinching, or twisting of the wrist? ☐ ☐ ☐

   Comments: _____

   _____

5. Is the force required to activate the controls no greater than 5 lbf? ☐ ☐ ☐

   Comments: _____

   _____

P. Lavatories

1. Is there a clear floor space 30 by 48 inches in front of a lavatory to allow a forward approach? ☐ ☐ ☐

   Comments: _____

   _____

2. Does the clear floor space adjoin or overlap an accessible route and extend a maximum of nineteen inches underneath the lavatory?

☐ ☐ ☐

Comments: _____

_____

3. Are lavatories mounted with the rim or counter surface no more than thirty-four inches above the finish floor?

☐ ☐ ☐

Comments: _____

_____

4. Is there a clearance space under the lavatory twenty-nine inches above the finished floor to the bottom of the lavatory apron?

☐ ☐ ☐

Comments: _____

_____

5. Is the knee clearance under the lavatory a minimum of 8 inches and the toe clearance 6 inches? The total distance from the front lip of the lavatory to the wall should be a minimum of 17 inches.

☐ ☐ ☐

Comments: _____

_____

6. Is the maximum depth of the sink 6½ inches?

☐ ☐ ☐

Comments: _____

_____

7. Are hot water and drain pipes either insulated or enclosed to protect against contact?

☐ ☐ ☐

Comments: _____

_____

8. Are faucets operable with one hand? Lever-operated, push type, and electronically controlled mechanisms are examples of acceptable design.

☐ ☐ ☐

Comments: _____

_____

9. If self-closing valves are used in faucets, do they remain open for at least ten seconds?

☐ ☐ ☐

Comments: _____

_____

10. Can faucets be operated with no more than 5 lbf?

☐ ☐ ☐

Comments: _____

_____

11. Are mirrors mounted with the bottom edge of the reflective surface no more than forty inches above the floor?

☐ ☐ ☐

Comments: _____

_____

Q. Handrails and Grab Bars

1. Are all handrails and grab bars accessible?

   Comments: _____

   _____

2. Are handrails installed along both sides of ramp segments? The inside handrail on switchback or dogleg ramps shall always be continuous.

   Comments: _____

   _____

3. If handrails are not continuous, do they extend at least twelve inches beyond the top and bottom of the ramp segment and are they parallel with the floor or ground surface?

   Comments: _____

   _____

4. Are the handrail gripping surfaces mounted between 35 and 38 inches above ramp surfaces?

   Comments: _____

   _____

5. Is the clear space between the handrail and the wall 1–1½ inches?

   Comments: _____

   _____

6. Are handrails and grab bars between 1.25 and 1.5 inches in diameter?

   Comments: _____

   _____

7. Are gripping surfaces continuous?

   Comments: _____

   _____

8. Are the ends of handrails either rounded or returned smoothly to floor, wall, or post?

   Comments: _____

   _____

9. Are handrails and grab bars secure in their fittings? They should not rotate within their fittings.

   Comments: _____

   _____

10. Are grab bars placed appropriately and where required, adjacent to toilets, toilet stalls, and urinals?

    Comments: _____

    _____

11. Is there a grab bar on the wall closest to the toilet, 40 inches in length, 33 to 36 inches from the floor, and 12 inches from the back wall? ☐ ☐ ☐

Comments: _____

_____

12. Is there at least 1.5 inches of space between grab bars and the wall? ☐ ☐ ☐

Comments: _____

_____

13. Are grab bars capable of resisting a force of 250 lbf? ☐ ☐ ☐

Comments: _____

_____

14. Are grab bars free of sharp, abrasive, or protruding elements? ☐ ☐ ☐

Comments: _____

_____

R. Controls and Operating Mechanisms

1. Are all controls and operating mechanisms accessible? ☐ ☐ ☐

Comments: _____

_____

2. Is there clear floor space at least 30 by 48 inches provided in front of the controls, dispensers, receptacles, and other operable equipment mechanisms to allow a forward approach by a person in a wheelchair, and is the clear space on an accessible route? ☐ ☐ ☐

Comments: _____

_____

3. Is the highest operable part of the controls, dispensers, receptacles, and other operable equipment placed within a maximum of forty-eight inches? ☐ ☐ ☐

Comments: _____

_____

4. Are electrical and communications system receptacles on walls mounted no less than fifteen inches above the floor? ☐ ☐ ☐

Comments: _____

_____

5. Are controls and operating mechanisms operable with one hand, and don't require tight grasping, pinching, or twisting the wrist? ☐ ☐ ☐

Comments: _____

_____

6. Is the force required to activate controls no greater than 5 lbf? ☐ ☐ ☐

Comments: _____

_____

S. Alarms

1. Are all alarm systems accessible? ☐ ☐ ☐

    Comments: _____

    _____

2. At a minimum, are visual signal appliances provided in buildings in restrooms, and
    any other general usage areas (meeting rooms, hallways, lobbies, and any other
    area for common use)? ☐ ☐ ☐

    Comments: _____

    _____

3. If provided, do audible emergency alarms produce a sound that exceeds the
    prevailing equivalent sound in the room or space by at least 15 dBA, or exceeds any
    maximum sound level with duration of 60 seconds by 5 dBA? The decibel (dB) is
    used to measure sound level, and dBA is a "weighting filter" to measure the sound
    pressure level in units of dBA. ☐ ☐ ☐

    Comments: _____

    _____

4. Does the sound level of the emergency alarm not exceed more than 110 dB at the
    minimum hearing distance from the audible appliance? ☐ ☐ ☐

    Comments: _____

    _____

5. If there is an emergency warning system (fire alarms), are visual signal appliances
    provided in restrooms and other general usage areas (e.g., meeting rooms, hallways,
    lobbies, and other areas) for common use? ☐ ☐ ☐

    Comments: _____

    _____

6. Are visible alarms located within the space they serve so that the signal is visible
    from within the space? ☐ ☐ ☐

    Comments: _____

    _____

7. Is the visual alarm system integrated into the building or facility alarm system? ☐ ☐ ☐

    Comments: _____

    _____

8. Do the visual alarm signals have the following minimum photometric and location
    features?

    a) A xenon strobe-type lamp or equivalent? ☐ ☐ ☐
    b) A color that shall be clear or nominal white (i.e., unfiltered or clear filtered
        white light)? ☐ ☐ ☐
    c) A maximum pulse duration of two tenths of one second (0.2 sec) with a
        maximum duty cycle of 40 percent? The pulse duration is defined as the time
        interval between initial and final points of 10 percent of the maximum signal. ☐ ☐ ☐

d) A flash rate of a minimum of 1 Hz and a maximum of 3 Hz. ☐ ☐ ☐

e) The appliance shall be placed 80 inches above the highest floor level within the space or 6 inches below the ceiling, whichever is lower. ☐ ☐ ☐

Comments: _____

_____

In general, no place in any room or space shall be more than fifty feet from the signal. In large rooms and spaces exceeding 100 ft (30 m) across, without obstructions 6 ft (2 m) above the finish floor, such as auditoriums, devices may be placed around the perimeter, spaced a maximum 100 ft (30 m) apart, in lieu of suspending appliances from the ceiling.

T. Signage

1. Do all signs designating permanent rooms and spaces in the building comply with the ADA's "Accessibility Guidelines for Buildings and Facilities" (ADAAG)? ☐ ☐ ☐

   Comments: _____

   _____

2. Is large, clearly printed signage provided to identify all areas and functions in the library for the deaf and visually impaired? ☐ ☐ ☐

   Comments: _____

   _____

3. Does directional and informational signage about functional spaces in the building comply with ADAAG? ☐ ☐ ☐

   Comments: _____

   _____

4. Do all accessible elements (i.e., entrance doors, restrooms, water fountains, and parking spaces) display the international symbol of accessibility? ☐ ☐ ☐

   Comments: _____

   _____

5. Are the signs placed perpendicular to the route of travel? ☐ ☐ ☐

   Comments: _____

   _____

6. Can permanent signs be approached without encountering a protruding object or standing within the area of a swing door? ☐ ☐ ☐

   Comments: _____

   _____

7. Do letters and numbers on signs have a width-to-height ratio between 3:5 and 1:1 and a stroke-width-to-height ratio between 1:5 and 1:10? ☐ ☐ ☐

   Comments: _____

   _____

8. Are characters and numbers on signs sized according to the viewing distance from which they are to be read? The minimum heights are measured using an uppercase X. Lowercase characters are permitted.

   □ □ □

   Comments: _____
   _____

9. Are letters and numbers raised 1/32 inch in uppercase san serif and accompanied with Grade 2 Braille? Raised characters shall be at least 5/8 inches high, but no higher than 2 inches.

   □ □ □

   Comments: _____
   _____

10. Are pictograms accompanied by the equivalent verbal description placed directly below the pictogram?

    □ □ □

    Comments: _____
    _____

11. Is the border dimension of the pictogram six inches minimum in height?

    □ □ □

    Comments: _____
    _____

12. Are the characters and backgrounds of permanent signs constructed with a matte, non-glare, eggshell-colored, or some other non-glare finish?

    □ □ □

    Comments: _____
    _____

13. Do characters and symbols contrast with their background—either light characters on a dark background or dark characters on a light background?

    □ □ □

    Comments: _____
    _____

14. Are permanent signs for rooms and spaces installed on the wall adjacent to the latch side of the door and mounted at sixty inches above the finish floor to the centerline of the sign?

    □ □ □

    Comments: _____
    _____

15. Is the mounting location for the signage above such that a person may approach within three inches of signage without encountering protruding objects or standing within the swing of a door?

    □ □ □

    Comments: _____
    _____

16. Are the following international symbols of accessibility displayed? (ADA "Standards for Accessible Design" have illustrations of the signs.)

    a) Volume control telephones are required to have a volume control identified by a sign containing a depiction of a telephone handset with radiating sound waves.   □ □ □

b) Text telephones are required to be identified by directional signage indicating the location of the nearest text telephone placed adjacent to all banks of telephones, which do not contain a text telephone. The sign must display the International TDD symbol. ☐ ☐ ☐

c) Assistive listening systems require a sign that identifies the availability of such systems and includes the international symbol of access for hearing loss. ☐ ☐ ☐

Comments: _____

_____

U. Telephones

1. Are public telephones identified by the international symbol of accessibility? ☐ ☐ ☐

   Comments: _____

   _____

2. Are public telephones accessible by providing a clear floor or ground space at least 30 by 48 inches that allows either a forward of parallel approach by a person using a wheelchair? ☐ ☐ ☐

   Comments: _____

   _____

3. Do bases, enclosures, and fixed seats not impede approaches to a telephone by those who use wheelchairs? ☐ ☐ ☐

   Comments: _____

   _____

4. Is the highest operable part of the telephone placed at accessible heights (forty-eight inches) for wheelchair users? ☐ ☐ ☐

   Comments: _____

   _____

5. Do hearing aid-compatible and volume control telephones:

   a) Have volume controls capable of a minimum of 12 dbA above normal? ☐ ☐ ☐
   b) Have push-button controls? ☐ ☐ ☐

   Comments: _____

   _____

6. Is the cord length from the telephone to the handset a minimum of at least twenty-nine inches? ☐ ☐ ☐

   Comments: _____

   _____

7. If telephone books are provided, are they located within the 48-inch accessible height? ☐ ☐ ☐

   Comments: _____

   _____

	YES	NO	N/A

8. If a text telephone is used with a pay telephone, are they permanently affixed within or adjacent to the telephone enclosure? ☐ ☐ ☐

    Comments: _____

    _____

9. Are pay telephones designed to accommodate a portable text telephone equipped with a shelf and an electrical outlet within or adjacent to the telephone enclosure? ☐ ☐ ☐

    Comments: _____

    _____

10. Is the telephone handset capable of being placed flush on the surface of the shelf? ☐ ☐ ☐

    Comments: _____

    _____

11. If pay telephones are not provided with the above accessible features, is an equivalent facilitation system provided at a service desk whenever the library is open? ☐ ☐ ☐

    Comments: _____

    _____

V. Fixed or Built-in Seating and Tables

1. Do at least 5 percent, or a minimum of one of each element of fixed seating, tables, or study carrels, have access for people with disabilities? ☐ ☐ ☐

    Comments: _____

    _____

2. Is there clear space for people in wheelchairs in front of fixed tables and study carrels of at least 30 by 48 inches in front of the furniture? This allows either a forward or parallel approach by a person using a wheelchair. ☐ ☐ ☐

    Comments: _____

    _____

3. Is there a knee clearance of 27 inches high, 30 inches wide, and 19 inches deep for people in wheelchairs in front of fixed tables and study carrels? ☐ ☐ ☐

    Comments: _____

    _____

4. Are the tops of tables and study carrels counters between 28 to 34 inches above the finish floor or ground? ☐ ☐ ☐

    Comments: _____

    _____

5. Do circulation, reference, and other service desks meet code requirements designated for built-in table and counters?

    a) Is the height of the countertop between 28 and 34 inches above the finished floor? For the circulation desk, the 34-inch dimension is preferred. ☐ ☐ ☐

	YES	NO	N/A

b) Is knee clearance at least 27 inches high? For the staff side of the desk, there should be a 19-inch-deep knee clearance space. ☐ ☐ ☐

c) Is the minimum clear area for a person in a wheelchair approaching the desk thirty-six inches wide? ☐ ☐ ☐

Comments: _____

_____

6. Does at least one lane at each checkout area meet the requirements for an accessible route? ☐ ☐ ☐

Comments: _____

_____

7. Is the minimum clear aisle space at card catalogs, OPAC desks, and magazine displays 36 inches, with a maximum reach height of 48 inches? ☐ ☐ ☐

Comments: _____

_____

8. Are aisles for bookstacks (storage of library materials) a minimum of 36 inches wide (42 inches preferred)? ☐ ☐ ☐

Comments: _____

_____

9. Are aisles for bookshelves (commonly used publications such as current periodicals and frequently used reference materials) forty-four inches minimum? ☐ ☐ ☐

Comments: _____

_____

10. Is assistance available to retrieve books off shelves? If so there is no height limit. If assistance is not available, the topmost shelf height is limited to fifty-four inches. ☐ ☐ ☐

Comments: _____

_____

W. Assembly Areas

1. Are wheelchair locations an integral part of any library fixed-seating plan in meeting and assembly rooms so as to provide people with physical disabilities a line of sight comparable to those members of the general public using the facility? ☐ ☐ ☐

Comments: _____

_____

2. Is the meeting room or assembly area served by an accessible route? ☐ ☐ ☐

Comments: _____

_____

	YES	NO	N/A

3. Does the facility observe the following capacity for wheelchair locations? ☐ ☐ ☐

Room Capacity	Minimum Accessible Spaces			
a) 4 to 25	1	☐	☐	☐
b) 26 to 50	2	☐	☐	☐
c) 51 to 300	4	☐	☐	☐
d) 301 to 500	6	☐	☐	☐
e) Over 500	6, plus 1 additional space for each seating capacity increase of 100	☐	☐	☐

When the seating capacity exceeds 300, wheelchair spaces shall be provided in more than one location.

Comments: _____

_____

4. Does each wheelchair location provide minimum clear ground or floor space 36 inches wide by 48 inches for forward or rear access and 66 inches minimum for side access? ☐ ☐ ☐

Comments: _____

_____

5. Is the floor surface in the meeting room or assembly room stable, firm, and slip-resistant? ☐ ☐ ☐

Comments: _____

_____

6. Does the meeting room provide an assistive listening system to augment standard public address and audio systems? ☐ ☐ ☐

Comments: _____

_____

7. If the meeting room has fixed seating, is the assistive listening system located within a 50-foot viewing distance of the stage or front area? ☐ ☐ ☐

Comments: _____

_____

8. Does the signage include the international symbol of access for the hearing-impaired to notify patrons of the availability of a listening system? ☐ ☐ ☐

Comments: _____

_____

X. Building Assistance Facilities

1. Is there a designated rescue assistance area in the facility? ☐ ☐ ☐

Comments: _____

_____

	YES	NO	N/A

2. Are there designated emergency routes in the facility? ☐ ☐ ☐

   Comments: _____

   _____

3. Are these routes easily identified? ☐ ☐ ☐

   Comments: _____

   _____

4. Are there signs to guide users in case of emergency? ☐ ☐ ☐

   Comments: _____

   _____

5. Are the signs illuminated? ☐ ☐ ☐

   Comments: _____

   _____

6. Do these signs point the way to the rescue assistance area? ☐ ☐ ☐

   Comments: _____

   _____

7. Is the rescue assistance area enclosed, smoke-proof, and vented to the outside? ☐ ☐ ☐

   Comments: _____

   _____

8. Is the rescue assistance area separated from the building interior by at least one fire-resistant door? ☐ ☐ ☐

   Comments: _____

   _____

9. Does the rescue assistance area provide at least two accessible 30-by-48-inch wheelchair spaces which do not encroach on the width of any required exit route? ☐ ☐ ☐

   Comments: _____

   _____

10. Is there a two-way communication system between the primary entrance and the rescue assistance area? ☐ ☐ ☐

    Comments: _____

    _____

Y. Service Animals

1. Are dogs allowed in the library? This is a policy question determined by the library administration. However, under the ADA, service animals are allowed in libraries. Service animals are defined as dogs that are individually trained to do work or perform tasks for people with disabilities. Examples of such work or tasks include guiding people who are blind, alerting people who are deaf, pulling a wheelchair, alerting and protecting a person who is having a seizure, reminding a person with

| | YES | NO | N/A |

mental illness to take prescribed medications, calming a person with post-traumatic stress disorder during an anxiety attack, or performing other duties.   ☐ ☐ ☐

Comments: _____

_____

2. Can other animals be service animals? Since March 15, 2011, only dogs are recognized as service animals under titles II and III of the ADA.   ☐ ☐ ☐

Comments: _____

_____

3. How are service dogs controlled? Under the ADA, service animals must be harnessed, leashed, or tethered, unless these devices interfere with the service animal's work or the individual's disability prevents using these devices. In that case, the individual must maintain control of the animal through voice, signal, or other effective controls.   ☐ ☐ ☐

Comments: _____

_____

4. Are all dogs service dogs? What questions can staff ask to determine if the dog is a service dog?

   a) Is the dog a service animal required because of a disability?   ☐ ☐ ☐
   b) What work or task has the dog been trained to perform?   ☐ ☐ ☐

   Comments: _____

   _____

5. What questions are not allowed to be asked about the dog and its owner?

   a) What is the person's disability?   ☐ ☐ ☐
   b) Does the owner have medical documentation of their disability?   ☐ ☐ ☐
   c) Does the owner have a special identification card or training documentation for the dog?   ☐ ☐ ☐
   d) Ask the dog owner to demonstrate the dog's ability to perform the work or task?   ☐ ☐ ☐

   Comments: _____

   _____

6. Can a service dog be prohibited if other library users and/or staff are allergic or afraid of the dog? Allergies and fear of dogs are not valid reasons for denying access or refusing service to people using service animals. When a person who is allergic to dog dander and a person who uses a service animal must spend time in the same room, they both should be accommodated by assigning them, if possible, to different locations within the room or different rooms in the facility.   ☐ ☐ ☐

   Comments: _____

   _____

7. When can a service dog be removed from the library? If the dog is out of control and the handler does not take effective action to control it or if the dog is not housebroken, the dog may be removed from the library.   ☐ ☐ ☐

   Comments: _____

   _____

8. Can emotional support dogs be allowed in the library? Emotional support dogs help individuals with emotional problems by providing comfort and support and are allowed in libraries under the Fair Housing Amendments Act. ☐ ☐ ☐

    Comments: _____

    _____

9. What is a companion animal and is it allowed in the library? A dog, cat, or other pet that provides health benefits to a person. Library policy may allow the companion animal into the library but access is not mandated for the animal unless it is a service dog. Only dogs are allowed to be service dogs under the ADA. ☐ ☐ ☐

    Comments: _____

    _____

Telecommunications, Electrical, and Miscellaneous Equipment

Some of the material below is based on the ANSI/TIA/EIA-568-C Standard and the ANSI/TIA/EIA-569-C Standard Two good sources for information are Aniexter (https://www.anixter.com/en_us.html) and Hubbel-Premise (www.argo-contar.com/download/passive/ANSI-TIA_Standards.pdf).

A. General Considerations

1. Are electrical power outlets and data outlets planned for all areas where electronic workstations will be planned? ☐ ☐ ☐

   Comments: _____

   _____

2. Is a wireless system planned with wireless access points? Wireless access points (AP or WAP) are specially configured nodes on wireless local area networks (WLAN). Access points act as a central transmitter and receiver of WLAN radio signals. ☐ ☐ ☐

   Comments: _____

   _____

3. Are workstations located and grouped to enhance noise control and privacy? ☐ ☐ ☐

   Comments: _____

   _____

4. Has equipment been selected with quiet operation in mind? ☐ ☐ ☐

   Comments: _____

   _____

B. Entrance Facility

1. Does the building have a telecommunications entrance facility? The Entrance
   Facility (EF) is located where the access provider and inter-building network
   cables enter the building. Outside plant cables, typically from underground, are
   terminated inside the entrance facility. This location is known as the demarcation
   point—the transition from access provider to customer-owned cable. It is the point
   at which outside cabling interfaces with the interior building backbone cabling. ☐ ☐ ☐

   Comments: _____
   _____

2. Is there an equipment room (essentially a large telecommunications closet) that
   houses the main distribution frame, private branch exchange (PBX), secondary
   voltage protection, and so on? The equipment room is often appended to the entrance
   facilities or a computer room to allow shared air conditioning, security, fire control,
   lighting, and limited access. ☐ ☐ ☐

   Comments: _____
   _____

3. Does the entrance facility meet all of the electrical, fire, building, and
   municipal codes? ☐ ☐ ☐

   Comments: _____
   _____

4. Does the entrance facility share other functions, including fire and security alarms,
   closed-circuit television (CCTV), cable television (CATV), PBX, and so on? ☐ ☐ ☐

   Comments: _____
   _____

5. Is the room in an area that is not subject to floods? ☐ ☐ ☐

   Comments: _____
   _____

6. Is the building entrance facility a locked, dedicated, and enclosed room with a
   plywood termination field provided on two walls? (The plywood should be 3/4 inch,
   with dimensions of 8 feet high × 39 inches wide.) ☐ ☐ ☐

   Comments: _____
   _____

7. Does the room have at least 150 square feet of floor space? The rule of thumb is to
   provide 0.75 square feet of equipment room floor space for every 100 square feet of
   user workstation area. ☐ ☐ ☐

   Comments: _____
   _____

	YES	NO	N/A

8. Is the room located away from sources of electromagnetic interference (transformers, motors, induction heaters, theft detection systems, etc.) until interference is less than 3V/m (volt per meter-unit of electrical strength) across the frequency spectrum?   ☐ ☐ ☐

   Comments: _____
   _____

9. Are all surfaces treated to reduce dust, and walls and ceilings painted white or pastel to improve visibility?   ☐ ☐ ☐

   Comments: _____
   _____

10. Is there a single or double (36 inches × 80 inches) lockable door in order to limit access to the room?   ☐ ☐ ☐

    Comments: _____
    _____

11. Has piping, ductwork, mechanical equipment, power cabling, and unrelated storage been kept out of the equipment room?   ☐ ☐ ☐

    Comments: _____
    _____

12. Is the room maintained 24 hrs/day, 365 days/year, at a temperature of 64 degrees to 75 degrees F, 30 percent to 55 percent humidity, with positive pressure?   ☐ ☐ ☐

    Comments: _____
    _____

13. Is there a minimum of two dedicated 15A, 100 VAC duplex outlets on separate circuits?   ☐ ☐ ☐

    Comments: _____
    _____

14. Are there convenience duplex outlets placed at six-foot intervals around the perimeter of the room?   ☐ ☐ ☐

    Comments: _____
    _____

15. Has an emergency power system been considered?   ☐ ☐ ☐

    Comments: _____
    _____

C. Equipment Room

1. Is there an equipment room? The equipment room (ER) is a centralized space for housing the core electronic equipment, such as computer servers, routers, hubs, and so on. The backbone cabling originates from the ER, which serves the entire building. The ER may function as a telecommunications room (TR), and also may contain an entrance facility. Stringent electrical and environmental requirements apply to the design of an ER to provide a suitable operating environment for active network equipment. ☐ ☐ ☐

    Comments: _____

    _____

2. Is the ER supplied with non-switched, conditioned power with backup? ☐ ☐ ☐

    Comments: _____

    _____

3. Is the ER located away from mechanical rooms, electrical distribution panels, and wet or dirty areas? ☐ ☐ ☐

    a) Is there a minimum of one ER per building? ☐ ☐ ☐
    b) Is a suspended ceiling prohibited? ☐ ☐ ☐
    c) Is there dedicated, unswitched electrical power with backup and surge protection? ☐ ☐ ☐
    d) Is the room temperature and humidity controlled? ☐ ☐ ☐
    e) Is a 24-hour security system installed? ☐ ☐ ☐
    f) Is there a separate fire suppression system? ☐ ☐ ☐
    g) Are all surfaces treated to reduce dust, and walls and ceilings painted white or pastel to improve visibility? ☐ ☐ ☐
    h) Is carpeting excluded from the space? ☐ ☐ ☐
    i) Are there single or double (36 inches × 80 inches) lockable doors in order to limit access to the room? ☐ ☐ ☐
    j) Has piping, ductwork, mechanical equipment, power cabling, and unrelated storage been kept out of the equipment room? ☐ ☐ ☐
    k) Is the room maintained 24 hrs/day, 365 days/year, at a temperature of 64 degrees to 75 degrees F, 30 percent to 55 percent humidity, with positive pressure? ☐ ☐ ☐
    l) Is there a minimum of two dedicated 15A, 100 VAC duplex outlets on separate circuits? ☐ ☐ ☐
    m) Are there convenience duplex outlets placed at six-foot intervals around the perimeter of the room? ☐ ☐ ☐
    n) Has an emergency power system been considered? ☐ ☐ ☐

    Comments: _____

    _____

D. Telecommunications Room

1. Is there a telecommunications room (TR)? The telecommunications room is an enclosed space for management and termination of backbone and horizontal cross-connections. Most, if not all of the environmental and operational requirements of the EF and ER should apply to the TR.

   ☐ ☐ ☐

   Comments: _____

   _____

2. Is there a TR to provide the horizontal cabling to all of the work areas on a single floor of a building?

   ☐ ☐ ☐

   Comments: _____

   _____

3. If the ER serves more than 20,000 sq. ft. of floor space, has a second room been considered?

   ☐ ☐ ☐

   Comments: _____

   _____

4. If the equipment room is more than 300 feet to a service point, have additional telecommunications closets been included? (Recommended size, 10 feet × 11 feet for each 10,000-square-foot area served.)

   ☐ ☐ ☐

   Comments: _____

   _____

5. Is the TR isolated from EMI (electromagnetic interference)?

   ☐ ☐ ☐

   Comments: _____

   _____

6. Does the room provide proper grounding and lighting?

   ☐ ☐ ☐

   Comments: _____

   _____

E. Horizontal Pathways

Horizontal pathways extend between the telecommunications closet and the work area. A variety of generic pathway options are available.

1. Plenum ceiling? Cable bundles running from the telecom closet along j-hooks suspended above a plenum ceiling, fanning out once a work zone is reached, dropping through interior walls or support columns or raceways, and terminating at an information outlet (IO)?

   ☐ ☐ ☐

   Comments: _____

   _____

2. Under-floor duct? (Single or dual-level rectangular ducts embedded in greater than 2.5-inch-thick concrete flooring.)

    ☐ ☐ ☐

    Comments: _____

    _____

3. Flush duct? (Single-level rectangular duct embedded flush in greater than one-inch-thick concrete flooring.)

    ☐ ☐ ☐

    Comments: _____

    _____

4. Multichannel raceway? (Cellular raceway ducts capable of routing telecom and power cabling separately in greater than three-inch-thick reinforced concrete.)

    ☐ ☐ ☐

    Comments: _____

    _____

5. Cellular floor? (Preformed hollows, or steel-lined cellar, are provided in concrete, with header ducts from the telecom closet arranged at right angles to the cells.)

    ☐ ☐ ☐

    Comments: _____

    _____

6. Trench duct? (A wide, solid tray, sometimes divided into compartments and fitted with a flat top with gaskets along its entire length, is embedded flush with the concrete finish.)

    ☐ ☐ ☐

    Comments: _____

    _____

7. Access floor? (Modular floor panels supported by pedestals, are used in computer rooms and equipment rooms.)

    ☐ ☐ ☐

    Comments: _____

    _____

8. Conduit? (Is only used when outlet locations are permanent, device density low, and flexibility for future changes is not required.)

    ☐ ☐ ☐

    Comments: _____

    _____

9. Perimeter pathways? (This option includes surface, recessed, molding, and multichannel raceways).

    ☐ ☐ ☐

    Comments: _____

    _____

10. Is the maximum horizontal cable length 90 m (295 ft.), independent of media type?

    ☐ ☐ ☐

    Comments: _____

    _____

F. Cabling and Outlets

1. Is a star topology-structured cabling system used? In a star topology, each work-area telecommunications outlet is connected to a cross-connect in a telecommunications closet. All cables from a floor or area in the building therefore run back to one central point for administration. Each telecommunication closet must be star wired back to the equipment room for the building. ☐ ☐ ☐

   Comments: _____

   _____

2. Is the structured cabling system compatible with the type of media to be used? ☐ ☐ ☐

   Comments: _____

   _____

3. Based on the media to be transmitted, what cable alternatives have been selected:

   a) Category 5E Cable? The minimum requirement should be unshielded twisted pair (UTP) N4-pair, 24-gauge, 100 ohm copper cable? (Unshielded twisted pair cables closely resemble telephone cables but are enhanced for data communications to allow higher frequency transmissions. Category 5 cables and connection hardware are the minimum usually required. They are rated up to 100 MHz and are designed to handle any current copper-based application for voice, video, or data.) ☐ ☐ ☐

   b) Category 6 or 6A Cable? Category 6 is now the suggested standard. It is similar to Category 5e, except that it is made to a higher standard and provides performance of up to 250 MHz, more than double category 5 and 5e. ☐ ☐ ☐

   c) Single-mode and multimode optical fiber cables? (The highest performing structured cabling systems use fiber optics, and will be the choice of most libraries in the long run as fiber costs decline). ☐ ☐ ☐

   Comments: _____

   _____

4. Does each workstation have a minimum of two information outlet ports? ☐ ☐ ☐

   a) Is the first outlet (mandatory): 4-pair 100 UTP or ScTP cable and connector (Category 5e min. recommended)? ☐ ☐ ☐

   b) Is the second outlet 4-pair 100 UTP cable and connector? (min. Category 5e, Category 6 is recommended). ☐ ☐ ☐

   c) Have the following alternative cabling systems been considered?

      i) Is multimode optical fiber cabling, 2-fiber (or higher fiber count) used? ☐ ☐ ☐

      ii) Is single-mode optical fiber cabling, 2-fiber (or higher fiber count) used? ☐ ☐ ☐

   Comments: _____

   _____

5. Is every seat in the library considered as a workstation and equipped with telecommunications outlets? (A minimum requirement is one outlet port required for voice and the other for data.) ☐ ☐ ☐

   Comments: _____

   _____

G. Wireless

1. Will the library provide customers with a wireless network for access to the Internet and library databases? A wireless network uses radio waves to provide communication, and may be used by library customers to access the Internet and library materials through library wireless workstations, and/or through their own personal computers if they are equipped to receive wireless signals. ☐ ☐ ☐

   a) Wireless access (WAP or AP) uses a WiFi standard (also known as IEEE 802.11b, g, o, n or 802.11ac). Most users can simply bring their wireless-enabled laptop computer or other wireless device to the library and turn it on. The computer will automatically recognize the wireless network.

   b) Wireless access points located throughout the library communicate with customers' wireless devices almost anywhere in the building. When a customer's wireless network card senses a signal, a message appears on their screen indicating a wireless network is available. Before connection is allowed, customers should be asked if they abide by the Library's Internet Public Use Policy and Guidelines. Usually if they click that they agree, they may use the service immediately.

   c) Library wireless electronic workstations are usually connected at all times.

   d) Wireless connectivity may be slower than wired connections and there may be instances when the need for higher bandwidth makes using a wired connection preferable. However, for many applications, such as web browsing and e-mail, wireless connectivity should be sufficient.

   e) Some researchers have found that current wireless protocols can be compromised using certain available hacking tools. An intruder can "listen in" on wireless traffic, and library users should be made aware of this potential issue.

   Comments: _____

   _____

2. Will the library use an 802.11ac wireless network (aka 5G WiFi)? This is the latest on the market and offers up to 1733 Mbps of wireless connection speed on the 5Ghz frequency band. First introduced in 2012, the new standard can now be found in many routers. ☐ ☐ ☐

   Comments: _____

   _____

3. Will omni-directional antennas that provide 360-degree coverage be considered? This type of antenna will provide good horizontal and vertical coverage. ☐ ☐ ☐

   Comments: _____

   _____

4. Has care been taken in locating wireless access points or nodes in the library? Wireless telecommunications networks require significant effort in placing nodes or access points within the library to ensure proper wireless coverage for users in the library. Placement is both an art and science. The access points will tie back to switches via Ethernet cables. ☐ ☐ ☐

   Comments: _____

   _____

5. In placing WAP for users, have the following issues been considered:

   a) Design to give the best signal to the most users or coverage area, or pick a minimum and ensure that every user or area gets a minimum signal or better? ☐ ☐ ☐

   b) Does each potential AP have access to power and Ethernet cables? ☐ ☐ ☐

   c) Distance is key, since data rates go down as distance increases, so the minimum distance is based on the design objective above. ☐ ☐ ☐

   d) Take into account the maximum cable length limitations (100 meters) for the cable running from the Ethernet switch to the access point. If a 100-meter cable won't reach your preferred access point location, then think about moving the access point or possibly using a WLAN bridge or optical fiber to make the connection. ☐ ☐ ☐

   e) Signals are shared, so the more users per WAP, the smaller each user's share of the bandwidth will be. ☐ ☐ ☐

   f) The physical access point or node should be as far out of reach as possible—preferably on or above the ceiling (depending on construction). However, some users want a physical indication of nodes so that they may locate near the location. ☐ ☐ ☐

   Comments: _____

   _____

6. Has a radio frequency (RF) site survey been done before installing the access points? The site survey will spot potential sources of RF interference and provide a basis for determining the most effective installation locations for access points. ☐ ☐ ☐

   a) The ultimate goal of an RF site survey is to supply enough information to determine the number and placement of access points that provides adequate coverage throughout the building. ☐ ☐ ☐

   b) An RF site survey also detects the presence of interference coming from other sources that could degrade the performance of the wireless LAN. ☐ ☐ ☐

   Comments: _____

   _____

7. Have the following steps been taken in conducting the RF site survey:

   a) Plot potential locations for AP from a building floor plan and tentatively lay out AP on the plan? ☐ ☐ ☐

   b) Visually inspect the facility by walking through the facility before performing any tests to verify the accuracy of the facility diagram? This is a good time to note any potential barriers that may impact RF signals. ☐ ☐ ☐

   c) Identify user areas and plot them on the facility diagram? ☐ ☐ ☐

   d) Determine preliminary access point locations? By considering the location of wireless users and range estimations of the wireless LAN products you're using, approximate the locations of access points that will provide adequate coverage throughout the user areas and plan on some overlap. ☐ ☐ ☐

   e) Have mounting locations been considered that will provide the best signal for the AP? These might be vertical posts or supports above ceiling tiles. Be sure to recognize suitable locations for installing the access point, antenna, data cable, and power line. ☐ ☐ ☐

   f) Lay out the AP on the building plan? ☐ ☐ ☐

g) Using commercially available software, test the sites for data rate, signal strength, and signal quality? Install an access point at each preliminary location, and monitor the site survey software readings by walking varying distances away from the access point. ☐ ☐ ☐

h) If the site meets the data rate, signal strength, and signal quality requirements, install the AP? ☐ ☐ ☐

i) Document the final locations on the building plan? This is vital in future remodeling. ☐ ☐ ☐

Comments: _____

_____

8. Have access points been kept away from metal pillars, filing cabinets, ductwork or concrete walls, or other sources of interference (such as machinery) where possible? ☐ ☐ ☐

Comments: _____

_____

9. Have the appropriate access point antennae for the area been selected? ☐ ☐ ☐

Comments: _____

_____

10. Have the access points been placed to minimize cable distance from the telecommunications closets, if possible? ☐ ☐ ☐

Comments: _____

_____

11. Is there an overlap of coverage between nodes? A 20 percent overlap in coverage for seamless roaming is a good rule of thumb. ☐ ☐ ☐

Comments: _____

_____

12. Is it possible to adjust node power to reduce or increase the signal to optimize the overlap instead of moving WAP physically closer or farther apart? ☐ ☐ ☐

Comments: _____

_____

H. Workstation Connections

1. What types of connections will be used in connecting the workstation equipment with the wired and/or wireless network? The work area components extend from the telecommunications outlet to each workstation. ☐ ☐ ☐

Comments: _____

_____

2. Is the maximum length for work area cords less than five meters? ☐ ☐ ☐

Comments: _____

_____

3. Is the work area wiring designed to be relatively simple to interconnect so that moves, new equipment, and changes are easily managed? ☐ ☐ ☐

   Comments: _____

   _____

4. Are there patch cables, modular cords, PC adapter cables, fiber jumpers, and so on? ☐ ☐ ☐

   Comments: _____

   _____

5. Are adapters, baluns, and so on external to the telecommunications outlets? ☐ ☐ ☐

   Comments: _____

   _____

I. Workstation Equipment

1. Do the workstations have:

   a) Hidden wiring? ☐ ☐ ☐
   b) All necessary connection outlets to connect to the library's network? ☐ ☐ ☐
   c) Adequate workspace? ☐ ☐ ☐
   d) Space for printers, paper, and supplies? ☐ ☐ ☐
   e) Back panels to hide connections and wires from customers? ☐ ☐ ☐

   Comments: _____

   _____

2. Will printers be available at every workstation, or will some or all printers print to a network printer? ☐ ☐ ☐

   Comments: _____

   _____

3. Are impact printers controlled? ☐ ☐ ☐

   Comments: _____

   _____

4. Do electronic workstations provide applications for word processing, spreadsheets, media, communications, and other applications? ☐ ☐ ☐

   Comments: _____

   _____

5. If there are not enough electronic workstations to meet peak demand, is there a system in place to allocate use? ☐ ☐ ☐

   Comments: _____

   _____

J. Telephone System

1. Is there a central telephone system?

   □ □ □

   Comments: _____

   _____

2. Is a switchboard operator required?

   □ □ □

   Comments: _____

   _____

3. If so, is there adequate space for operators to do other work when not answering the phone?

   □ □ □

   Comments: _____

   _____

4. Are telephones hardwired?

   □ □ □

   Comments: _____

   _____

5. Does the telephone system allow for a staff member's business line to be transferred to their smartphone?

   □ □ □

   Comments: _____

   _____

6. Are alternative long-distance vendors used?

   □ □ □

   Comments: _____

   _____

7. Are incoming lines sufficient in number and quality?

   □ □ □

   Comments: _____

   _____

8. Are public telephones located to allow for convenient use while preventing disturbance to other customers?

   □ □ □

   Comments: _____

   _____

9. Are the public telephones set up for outgoing calls only?

   □ □ □

   Comments: _____

   _____

10. Are coin-changing machines available near the telephones?

    □ □ □

    Comments: _____

    _____

	YES	NO	N/A

11. Is the library involved in a network with branches, campus locations, and/or other libraries via telecommunications and data transfer?   ☐ ☐ ☐

   Comments: _____

   _____

12. Will Voice over Internet Protocol (VoIP) be used? VoIP is a technology that allows users to make voice calls using a broadband Internet connection instead of a regular (or analog) phone line. Some VoIP services may only allow calls to other people using the same service, but others may allow calls to anyone who has a telephone number—including local, long distance, mobile, and international numbers.   ☐ ☐ ☐

   Comments: _____

   _____

K. Miscellaneous Electrical Equipment

1. Are video monitors and television sets available for public use?   ☐ ☐ ☐

   Comments: _____

   _____

2. Are there audio players (DVD, MP3, tablets, etc.) available for use by the library's customers?   ☐ ☐ ☐

   Comments: _____

   _____

3. Are the video sets controlled by staff?   ☐ ☐ ☐

   Comments: _____

   _____

4. Is there provision for large-screen television viewing in meeting or conference rooms?   ☐ ☐ ☐

   Comments: _____

   _____

5. Is there access to cable TV?   ☐ ☐ ☐

   Comments: _____

   _____

6. Is there access to campus, private, government, or local broadcasts?   ☐ ☐ ☐

   Comments: _____

   _____

7. Are there teleconferencing and distance-learning facilities?   ☐ ☐ ☐

   Comments: _____

   _____

	YES	NO	N/A

8. Is microwave transmission/reception used? ☐ ☐ ☐

Comments: _____

_____

9. Is there a public-address system? ☐ ☐ ☐

Comments: _____

_____

10. Is there a video surveillance system? ☐ ☐ ☐

Comments: _____

_____

L. Electrical Power

1. Is there sufficient power distribution throughout the entire facility? ☐ ☐ ☐

Comments: _____

_____

2. Is it "clean power," of high quality, and reliable? ☐ ☐ ☐

Comments: _____

_____

3. Does the library generate its own power using solar, wind, and so on? ☐ ☐ ☐

Comments: _____

_____

4. Is there a backup power system in place? ☐ ☐ ☐

Comments: _____

_____

5. Does the system provide for future needs? ☐ ☐ ☐

Comments: _____

_____

6. Is all wiring easily accessible (raised floors, flat wire, grids under carpet, conduits
    above dropped ceilings or in columns)? ☐ ☐ ☐

Comments: _____

_____

7. Is surge protection available where needed? ☐ ☐ ☐

Comments: _____

_____

8. Is voltage regulated at the building feed? ☐ ☐ ☐

Comments: _____

_____

	YES	NO	N/A

9. Is voltage regulated at each floor box? ☐ ☐ ☐

   Comments: _____

   _____

10. Are dedicated lines available for equipment that requires them (terminals, photocopiers, etc.)? ☐ ☐ ☐

    Comments: _____

    _____

11. Are cords and cables protected and out of sight? ☐ ☐ ☐

    Comments: _____

    _____

12. Does each staff workstation have 3 to 5 duplex outlets? ☐ ☐ ☐

    Comments: _____

    _____

13. Are there convenience outlets at frequent intervals throughout the building? ☐ ☐ ☐

    Comments: _____

    _____

14. Do outlets have electrical and data/telephone capabilities? ☐ ☐ ☐

    Comments: _____

    _____

15. Are there specialized wiring arrangements (e.g., wall-mounted power strips or ceiling outlets) for areas such as teleconference, automated demonstration, and computing rooms? ☐ ☐ ☐

    Comments: _____

    _____

16. Are public workstations/carrels provided with power and data ports? ☐ ☐ ☐

    Comments: _____

    _____

17. Is there a user fee for using the library's power? ☐ ☐ ☐

    Comments: _____

    _____

Interior Design and Finishes

A. Service Desks

1. Whom does the service desk serve?

    a) Faculty?   ☐ ☐ ☐
    b) Students?   ☐ ☐ ☐
    c) Adults?   ☐ ☐ ☐
    d) Children?   ☐ ☐ ☐

    Comments: _____

    _____

2. What type of service desks are required:

    a) Control or security desk?   ☐ ☐ ☐
    b) Directional or information desk?   ☐ ☐ ☐
    c) Circulation or charge desk?   ☐ ☐ ☐
    d) Children's desk?   ☐ ☐ ☐
    e) Young adult desk?   ☐ ☐ ☐
    f) Media center desk?   ☐ ☐ ☐
    g) Call, reserve, or delivery desk?   ☐ ☐ ☐
    h) Reference desk?   ☐ ☐ ☐
    i) Reference or bibliographic consultation center?   ☐ ☐ ☐

    Comments: _____

    _____

3. Has the library considered not having a service desk? Some libraries have "roving staff" to create a more user-friendly environment where library staff approach and greet customers rather than the other way around. This enables the library to exploit self-service technologies by providing self-service checkouts, computer terminals, and shopping center-style kiosks.

    □ □ □

    Comments: _____

    _____

4. Is the design of the desk area flexible, allowing possible future relocation, new technology, or even elimination of the desk?

    □ □ □

    Comments: _____

    _____

5. What kind of way-finding system and signs lead people to the service desks?

    □ □ □

    Comments: _____

    _____

6. Is the desk in a visible location so that it is obvious to people who need the services provided at the desk?

    □ □ □

    Comments: _____

    _____

7. Is the desk sized to accommodate all staff working at the desk, as well as their storage requirements?

    □ □ □

    Comments: _____

    _____

8. Are the desk and surrounding workspace designed to be ergonomically correct for staff and customers?

    □ □ □

    Comments: _____

    _____

9. Have customer self-service features been factored into the desk, such as self-check, electronic registration, and so on?

    □ □ □

    Comments: _____

    _____

10. Can conversations at the desk be conducted with a sense of privacy?

    □ □ □

    Comments: _____

    _____

11. How has noise from the service desk from conversations, equipment, phones, and so on been addressed so that nearby spaces are not disrupted?

    □ □ □

    Comments: _____

    _____

	YES	NO	N/A

12. Have openness and accessibility been maintained while protecting staff from potential aggressive customers? ☐ ☐ ☐

   Comments: _____

   _____

13. Are grommets, wire channels, and equipment shielding provided and designed to present a clean appearance? ☐ ☐ ☐

   Comments: _____

   _____

14. Have sufficient electrical outlets and data and telephone ports been provided for current use as well as anticipated use? ☐ ☐ ☐

   Comments: _____

   _____

15. Is the desk protected from direct sunlight, drafts, or other adverse environmental conditions? ☐ ☐ ☐

   Comments: _____

   _____

16. Are the desk finishes and materials highly durable and relatively vandal-proof? ☐ ☐ ☐

   Comments: _____

   _____

17. Can the desk surfaces and edges be easily cleaned? ☐ ☐ ☐

   Comments: _____

   _____

B. Plus-Friendly Spaces

More than one-third of adults and 17 percent of children may be classified as obese based on a data-brief from the Centers for Disease Control and Prevention's National Center for Health Statistics. Consequently, the library needs to be planned not only to meet ADA guidelines, but also to make it welcome to plus-sized users and staff.

A good reference for information about serving a larger population is the National Association to Advance Fat Acceptance (NAAFA): www.naafaonline.com/dev2/index.html. NAAFA is a nonprofit, all-volunteer, civil rights organization dedicated to protecting the rights and improving the quality of life for fat people.

Plus-friendly furniture should be mixed in with other furnishings so all the furniture has the same look and style.

Some of the considerations below may be now or will become required as new rules and regulations are mandated by ADA standards.

1. Is the staff trained and sensitive to the special needs and wants of bariatric patrons? ☐ ☐ ☐

   Comments: _____

   _____

	YES	NO	N/A

2. Are doorways larger than normal to accommodate oversized wheelchairs? ☐ ☐ ☐

   Comments: _____

   _____

3. Are public elevators large enough to accommodate larger people? ☐ ☐ ☐

   Comments: _____

   _____

4. In case of emergency, are staff or service elevators large enough to accommodate a stretcher or gurney as well as health care personnel? ☐ ☐ ☐

   Comments: _____

   _____

5. In order to reduce the need to walk around, can library information be made available by posting QR codes throughout the building? ☐ ☐ ☐

   Comments: _____

   _____

6. Will some task chairs support people weighing up to 500 lbs.? ☐ ☐ ☐

   Comments: _____

   _____

7. Are the seats and backs 20 percent larger than standard products? ☐ ☐ ☐

   Comments: _____

   _____

8. Are these chairs available without armrests? ☐ ☐ ☐

   Comments: _____

   _____

9. Are chairs available in a variety of widths (more than thirty-six inches recommended)? A search for bariatric chairs may provide some good choices. ☐ ☐ ☐

   Comments: _____

   _____

10. Are chairs and desks separate pieces? A chair can be moved to a comfortable distance from the desk. ☐ ☐ ☐

   Comments: _____

   _____

11. Are booths not affixed to the floor or wall so that they may be repositioned in relation to seats? ☐ ☐ ☐

   Comments: _____

   _____

12. Are the plus-sized seating areas distributed throughout the building? ☐ ☐ ☐

   Comments: _____

   _____

	YES	NO	N/A

13. Are step stools labeled with weight limits? ☐ ☐ ☐

   Comments: _____

   _____

14. Are there plus-sized seating areas for children and young adults? ☐ ☐ ☐

   Comments: _____

   _____

15. Do public auditoriums and theatres provide oversize seating? Generally, larger seats
   should blend into the project and not be apparent. Larger patrons must identify them-
   selves when reserving tickets or seating directions in order to reserve a larger seat. ☐ ☐ ☐

   Comments: _____

   _____

16. Do conference and/or seminar rooms have chairs properly sized and without arms? ☐ ☐ ☐

   Comments: _____

   _____

17. Do some of the toilet rooms have floor-mounted stainless steel toilets with a capacity
   of 500 pounds? ☐ ☐ ☐

   Comments: _____

   _____

18. Are these toilets higher than normal? ☐ ☐ ☐

   Comments: _____

   _____

19. Do the toilets have oversized seats? ☐ ☐ ☐

   Comments: _____

   _____

20. Is the toilet paper dispenser placed about 12–18 inches in front of the toilet rather
   than beside it? ☐ ☐ ☐

   Comments: _____

   _____

C. Seating

1. Is there variety in the types of seating? ☐ ☐ ☐

   Comments: _____

   _____

2. Is lounge seating modular or heavy enough to discourage casual rearrangement by
   customers, unless the library desires rearrangement? ☐ ☐ ☐

   Comments: _____

   _____

3. Is adequate and appropriate seating provided for varying tasks and areas:

    a) Staff work areas? ☐ ☐ ☐
    b) Public seating at tables and carrels? ☐ ☐ ☐
    c) Faculty and graduate carrels and rooms? ☐ ☐ ☐
    d) Information commons? ☐ ☐ ☐
    e) Reference areas? ☐ ☐ ☐
    f) Lounge areas? ☐ ☐ ☐
    g) Meeting rooms? ☐ ☐ ☐

    Comments: _____

    _____

4. Is seating appropriate for different ages? ☐ ☐ ☐

    Comments: _____

    _____

5. Are people (especially senior citizens) able to get in and out of chairs easily? ☐ ☐ ☐

    Comments: _____

    _____

6. Is seating comfortable for those areas where the library wants users to relax and
    read for an extended period? ☐ ☐ ☐

    Comments: _____

    _____

7. Is seating comfortable but conducive to quick turnover for those areas where you
    want users to leave after their work task is completed? (The two types of seating
    can be exemplified by the seating available in a fast-food restaurant versus that
    found in a fine restaurant.) ☐ ☐ ☐

    Comments: _____

    _____

8. Are chairs ergonomically correct? ☐ ☐ ☐

    Comments: _____

    _____

9. Is seating attractive and inviting? ☐ ☐ ☐

    Comments: _____

    _____

10. Is furniture free of projections that could snag clothing? ☐ ☐ ☐

    Comments: _____

    _____

11. Is furniture relatively free of sharp corners? ☐ ☐ ☐

    Comments: _____

    _____

12. Does seating take personal space into consideration to avoid psychological feelings of crowding? ☐ ☐ ☐

    Comments: _____

    _____

13. If the chair has arms, will the arms fit comfortably under work surfaces? ☐ ☐ ☐

    Comments: _____

    _____

14. Are footstools or ottomans provided? ☐ ☐ ☐

    Comments: _____

    _____

15. Is furniture designed for easy repair or replacement of parts? ☐ ☐ ☐

    Comments: _____

    _____

16. Is furniture constructed for user safety? ☐ ☐ ☐

    Comments: _____

    _____

17. Has the furniture been used successfully in similar library or other public situations for several years? ☐ ☐ ☐

    Comments: _____

    _____

18. Are performance data available to attest to the durability of the chair? ☐ ☐ ☐

    Comments: _____

    _____

19. Has seating been stress tested? ☐ ☐ ☐

    Comments: _____

    _____

20. Have chairs been tested for at least one day by a variety of users? This test of "100 bottoms" should allow a variety (sex, height, weight) of people to sit in and use the furniture for at least one day. ☐ ☐ ☐

    Comments: _____

    _____

21. Do chairs with casters move easily on carpet? ☐ ☐ ☐

    Comments: _____

    _____

22. Is lounge seating modular or heavy enough not to tip over? ☐ ☐ ☐

    Comments: _____

    _____

	YES	NO	N/A

23. Are fabrics sturdy and soil-resistant? ☐ ☐ ☐

    Comments: _____

    _____

24. Do the chair design and the kind of upholstery or finish used allow for easy cleaning? ☐ ☐ ☐

    Comments: _____

    _____

25. Can the chair be easily reupholstered or refinished? ☐ ☐ ☐

    Comments: _____

    _____

26. Is the fabric porous enough to "breathe" and able to absorb and evaporate moisture easily? ☐ ☐ ☐

    Comments: _____

    _____

27. Do lounge chairs with upholstered arms have arm covers to preserve their appearance? ☐ ☐ ☐

    Comments: _____

    _____

28. Are chairs designed so that the area under the chair can be easily reached by a
    vacuum cleaner? ☐ ☐ ☐

    Comments: _____

    _____

29. Does the supplier warranty the design and construction of the seats? ☐ ☐ ☐

    Comments: _____

    _____

30. What is the length of the warranty? ☐ ☐ ☐

    Comments: _____

    _____

D. Tables

1. Are the tables appropriate for the task intended? ☐ ☐ ☐

   Comments: _____

   _____

2. Are the tables durable and strong? ☐ ☐ ☐

   Comments: _____

   _____

3. Is the work surface material appropriate for the use anticipated? ☐ ☐ ☐

   Comments: _____

   _____

	YES	NO	N/A

4. Can the work surface be easily maintained? ☐ ☐ ☐

   Comments: _____

   _____

5. Can the work surface be easily refinished? ☐ ☐ ☐

   Comments: _____

   _____

6. Does the table have any needed accessories, such as task lighting, electrical outlets, and so on? ☐ ☐ ☐

   Comments: _____

   _____

7. Is there a mixture of circular tables (for socializing) and rectangular tables (better for work and concentration) on the floor? ☐ ☐ ☐

   Comments: _____

   _____

8. Are there enough carrels for individual studying? ☐ ☐ ☐

   Comments: _____

   _____

9. Does the supplier warranty the design and construction of the table? ☐ ☐ ☐

   Comments: _____

   _____

10. What is the length of the warranty? ☐ ☐ ☐

   Comments: _____

   _____

E. Lighting (Please also see the lighting section in chapter 12, "Building Systems")

1. Is the intensity of the general lighting sufficient for reading? ☐ ☐ ☐

   Comments: _____

   _____

2. Is the "task lighting" adequate for carrels, workstations, separate desks, lounge furniture, and shelving areas? ☐ ☐ ☐

   Comments: _____

   _____

3. In addition to general and task lighting, do certain areas of the library have special lighting? For example, do wall display areas have track lighting? ☐ ☐ ☐

   Comments: _____

   _____

	YES	NO	N/A

4. Are the lights used to highlight display cases and exhibits non-glaring? ☐ ☐ ☐

   Comments: _____

   _____

5. Is lighting adequate at the lower shelf areas in book stacks? (Lighting levels drop
   dramatically from the top to the bottom of book stacks.) ☐ ☐ ☐

   Comments: _____

   _____

6. Are light switches conveniently located? ☐ ☐ ☐

   Comments: _____

   _____

7. Can library staff control the switching of lights from a central control point or points? ☐ ☐ ☐

   Comments: _____

   _____

8. Is the lighting control system designed so that customers can't switch lights on and
   off in those areas where public control is not desirable? ☐ ☐ ☐

   Comments: _____

   _____

9. Do occupancy sensor timers control room lighting? Occupancy sensors detect
   indoor activity within a certain area. They provide convenience by turning lights on
   automatically when someone enters a room, and save energy by turning lights off
   soon after the last occupant has left the room. Occupancy sensors must be located
   where they will detect occupants or occupant activity in all parts of the room. ☐ ☐ ☐

   Comments: _____

   _____

10. Are timers used to turn on and off outdoor lights at specific times? ☐ ☐ ☐

    Comments: _____

    _____

11. Are photo sensors used to control outdoor lights? They may be designed to prevent
    outdoor lights from operating during daylight hours. Lights may be programmed to
    go off after a certain number of hours. ☐ ☐ ☐

    Comments: _____

    _____

F. Windows

1. Has the library considered the trade-off between the positive aspects of operable windows (natural light, fresh air, and pleasant vistas) vs. the negative factors (the possible waste of energy, the loss of outside walls as book stack areas, and the impact of uncontrolled sunlight on materials and readers)?   ☐ ☐ ☐

   Comments: _____
   _____

2. If the regional climate allows it, are windows operable to allow for natural cooling and ventilation?   ☐ ☐ ☐

   Comments: _____
   _____

3. If operable windows are used, how will staff control their use?   ☐ ☐ ☐

   Comments: _____
   _____

4. If windows can be opened, are they securable by the staff from the inside?   ☐ ☐ ☐

   Comments: _____
   _____

5. In those libraries where operable windows are not employed, are a limited number of windows operable to allow for maintenance and emergency situations?   ☐ ☐ ☐

   Comments: _____
   _____

6. Are some of the windows placed close to the ceiling to allow a higher intensity of light?   ☐ ☐ ☐

   Comments: _____
   _____

7. Are some of the windows placed at eye level, especially in reading areas and in areas occupied by the staff, for positive psychological effect?   ☐ ☐ ☐

   Comments: _____
   _____

8. Can windows be shaded on demand to prevent light from interfering with reading and other activities?   ☐ ☐ ☐

   Comments: _____
   _____

9. Are books stored away from direct sunlight to protect the bindings from fading and to prevent paper deterioration?   ☐ ☐ ☐

   Comments: _____
   _____

	YES	NO	N/A

10. If the regional climate suggests it, are windows double-glazed to allow for enhanced energy efficiency? ☐ ☐ ☐

    Comments: _____

_____

11. Are windows accessible for cleaning and maintenance? ☐ ☐ ☐

    Comments: _____

_____

12. Will special equipment be required to maintain and clean windows? ☐ ☐ ☐

    Comments: _____

_____

G. Flooring

1. Has the trade-off between types of floor coverings been considered by examining the:

    a) Original construction costs? ☐ ☐ ☐

    b) Total useful life of the floor covering? ☐ ☐ ☐

    c) Appropriateness of the floor covering for the area to be covered? Soiling potential, spills, noise, and so on. ☐ ☐ ☐

    d) Ease of maintenance? ☐ ☐ ☐

    e) Cost of maintenance? ☐ ☐ ☐

    f) Ease of replacement? ☐ ☐ ☐

    g) Cost of replacement? ☐ ☐ ☐

    Comments: _____

_____

2. Are special floor-covering materials or systems used at the entry and places of heavy traffic to prevent dirt, mud, slush, and water from being tracked onto the carpet? ☐ ☐ ☐

    Comments: _____

_____

3. Have carpet tiles or squares been considered for easy access to under-floor power systems as well as ease of replacement when damaged or soiled? ☐ ☐ ☐

    Comments: _____

_____

4. Is the carpet of first-class quality to ensure durability? ☐ ☐ ☐

    Comments: _____

_____

5. Does the carpet color conceal soiling and resist fading? ☐ ☐ ☐

    Comments: _____

_____

	YES	NO	N/A

6. Does flooring minimize noise and enhance building acoustics? ☐ ☐ ☐

   Comments: _____

   _____

7. Can book trucks be moved easily across the flooring? ☐ ☐ ☐

   Comments: _____

   _____

8. Is ceramic tile or a similar material used on the restroom floors for its sanitary
   appearance and ease of maintenance? ☐ ☐ ☐

   Comments: _____

   _____

9. If pavement tiles (stone, marble, or granite) are used in entryways and lobbies, are
   provisions made for safety since these have the potential to become very slippery
   when wet? ☐ ☐ ☐

   Comments: _____

   _____

10. Has concrete flooring, if left uncovered, been treated with filler and then painted to
    prevent dust from becoming troublesome? ☐ ☐ ☐

    Comments: _____

    _____

11. If wood floors are used, does the library's operating budget allow for the care needed
    to keep them in good condition? Hard species, such as red oak, will withstand wear
    and tear, while softer species, such as pine, tend to show scratches. ☐ ☐ ☐

    Comments: _____

    _____

H. Walls

1. Have "wet" interior walls been avoided as much as possible? (Wet walls are those
   that cannot be removed without demolishing them.) ☐ ☐ ☐

   Comments: _____

   _____

2. Are the wall finishes, coverings, and surfaces appropriate for the room's function? ☐ ☐ ☐

   Comments: _____

   _____

3. Will the care and selection of wall coverings result in years of added wear and
   minimum upkeep? ☐ ☐ ☐

   Comments: _____

   _____

	YES	NO	N/A

4. Are areas subject to soiling covered with a washable paint with a glossy finish? ☐ ☐ ☐

   Comments: _____

   _____

5. Is a matte or dull finish used where reflectivity is a concern? ☐ ☐ ☐

   Comments: _____

   _____

6. To add interest, are there special wall treatments such as stenciling, textured materials such as a woven fabric, or wood paneling? ☐ ☐ ☐

   Comments: _____

   _____

7. Have other materials such as brick and stone been used for wall coverings? ☐ ☐ ☐

   Comments: _____

   _____

8. Is ceramic tile used for the walls in the restrooms for ease of maintenance? ☐ ☐ ☐

   Comments: _____

   _____

9. If ceramic tile has been used to create decorative wall murals, has care been taken to minimize the acoustical impact of the hard surface? ☐ ☐ ☐

   Comments: _____

   _____

10. Have vinyl wall coverings been considered for areas of heavy use, including hallways and staircases? ☐ ☐ ☐

    Comments: _____

    _____

11. Have vinyl wall coverings with special sound-absorbing properties been considered for offices, workrooms, and conference rooms? ☐ ☐ ☐

    Comments: _____

    _____

12. Do the walls have tack board or other such surfaces so that they may be used for occasional displays? ☐ ☐ ☐

    Comments: _____

    _____

13. Are white boards used in rooms where meetings or informal discussions take place? ☐ ☐ ☐

    Comments: _____

    _____

14. Electronic display walls? ☐ ☐ ☐

    Comments: _____

    _____

I. Color

1. Have colors that may quickly become outdated been avoided? ☐ ☐ ☐

   Comments: _____

   _____

2. Has particular attention been given to the psychological effects of color on both users and staff? ☐ ☐ ☐

   Comments: _____

   _____

3. Has color been considered with respect to the function of the area? ☐ ☐ ☐

   Comments: _____

   _____

4. Has color been used to avoid an institutional (drab) aspect with respect to walls, book stacks, floors, and furniture? ☐ ☐ ☐

   Comments: _____

   _____

5. Do book stacks on different floors or areas utilize different colors for easy identification? ☐ ☐ ☐

   Comments: _____

   _____

6. Have standard paint colors (not mixed) been supplied by the manufacturer for easy, cost-effective maintenance and touch-ups? ☐ ☐ ☐

   Comments: _____

   _____

7. Will the upholstery colors selected disguise heavy and sometimes abusive use? ☐ ☐ ☐

   Comments: _____

   _____

J. Equipment List

Is the following equipment planned for use in the library? If so, are there adequate space, wiring, furniture, and staff available to support the equipment?

1. Online public access terminals? ☐ ☐ ☐

   Comments: _____

   _____

2. Public electronic workstations? ☐ ☐ ☐

   Comments: _____

   _____

	YES	NO	N/A

3. Public printers? ☐ ☐ ☐

   Comments: _____

   _____

4. Staff telephones with hold and transfer capabilities? ☐ ☐ ☐

   Comments: _____

   _____

5. Public telephones? ☐ ☐ ☐

   Comments: _____

   _____

6. Mobile two-way communication system? ☐ ☐ ☐

   Comments: _____

   _____

7. Answering machines or voice mail? ☐ ☐ ☐

   Comments: _____

   _____

8. Staff paging communication devices? ☐ ☐ ☐

   Comments: _____

   _____

9. Public address system? ☐ ☐ ☐

   Comments: _____

   _____

10. Tele-facsimile (fax) machines? ☐ ☐ ☐

    Comments: _____

    _____

11. Voice-synthesis reading machines? ☐ ☐ ☐

    Comments: _____

    _____

12. Public text telephones (TDDs)? ☐ ☐ ☐

    Comments: _____

    _____

13. Electric typewriters? ☐ ☐ ☐

    Comments: _____

    _____

14. Large-print typewriters? ☐ ☐ ☐

    Comments: _____

    _____

	YES	NO	N/A

15. Word processor workstations? ☐ ☐ ☐

Comments: _____

_____

16. Audio recorders/players? ☐ ☐ ☐

Comments: _____

_____

17. Video recorders/players? ☐ ☐ ☐

Comments: _____

_____

18. Tape duplicators? ☐ ☐ ☐

Comments: _____

_____

19. MP3 players ☐ ☐ ☐

Comments: _____

_____

20. Video disc players? ☐ ☐ ☐

Comments: _____

_____

21. Compact disc players? ☐ ☐ ☐

Comments: _____

_____

22. Record players? ☐ ☐ ☐

Comments: _____

_____

23. Audiocassette players? ☐ ☐ ☐

Comments: _____

_____

24. Game hardware? ☐ ☐ ☐

Comments: _____

_____

25. Headphones? ☐ ☐ ☐

Comments: _____

_____

26. Film projectors and screens? ☐ ☐ ☐

Comments: _____

_____

27. Video projectors? ☐ ☐ ☐

    Comments: _____

    _____

28. Slide projectors? ☐ ☐ ☐

    Comments: _____

    _____

29. Light table (for slides and/or tracing maps)? ☐ ☐ ☐

    Comments: _____

    _____

30. Overhead projectors? ☐ ☐ ☐

    Comments: _____

    _____

31. Microform readers? ☐ ☐ ☐

    Comments: _____

    _____

32. Microform readers/printers? ☐ ☐ ☐

    Comments: _____

    _____

33. Photocopiers? ☐ ☐ ☐

    Comments: _____

    _____

34. Card-operated photocopiers? ☐ ☐ ☐

    Comments: _____

    _____

35. Clocks strategically located and visible in all public places, as well as easily accessible
    or centrally controlled? ☐ ☐ ☐

    Comments: _____

    _____

36. Time clocks? ☐ ☐ ☐

    Comments: _____

    _____

37. Fire hoses? ☐ ☐ ☐

    Comments: _____

    _____

38. Fire extinguishers? ☐ ☐ ☐

    Comments: _____

    _____

	YES	NO	N/A

39. Emergency lights? □ □ □

Comments: _____
_____

40. Emergency power (generators)? □ □ □

Comments: _____
_____

41. Closed-circuit TV systems? □ □ □

Comments: _____
_____

42. Security mirrors? □ □ □

Comments: _____
_____

43. Emergency call system direct to police or security company? □ □ □

Comments: _____
_____

44. Emergency call buttons located at service desks and workrooms? □ □ □

Comments: _____
_____

45. Book trucks:
   a) Are they top quality with solid joints and pivoting wheels? □ □ □
   b) Do they roll smoothly and quietly on all floor surfaces? □ □ □
   c) Are they equipped with shelf height, depth, and slant to accommodate
       materials of various sizes? □ □ □
   d) Are there sufficient quantities of trucks in various sizes and configurations? □ □ □

Comments: _____
_____

46. Chalkboards/white boards? □ □ □

Comments: _____
_____

47. Bulletin boards □ □ □

Comments: _____
_____

48. Easels? □ □ □

Comments: _____
_____

49. Lecterns? □ □ □

Comments: _____
_____

	YES	NO	N/A

50. Display racks? ☐ ☐ ☐

Comments: _____

_____

51. Marketing fixtures? ☐ ☐ ☐

Comments: _____

_____

52. Bookends in appropriate sizes and shapes? ☐ ☐ ☐

Comments: _____

_____

53. Pencil sharpeners? ☐ ☐ ☐

Comments: _____

_____

54. Pencil dispensers? ☐ ☐ ☐

Comments: _____

_____

55. Customer vending machines for supplies? ☐ ☐ ☐

Comments: _____

_____

56. Filing cabinets? ☐ ☐ ☐

Comments: _____

_____

57. Electric staplers? ☐ ☐ ☐

Comments: _____

_____

58. Hole punches? ☐ ☐ ☐

Comments: _____

_____

59. Paper cutters? ☐ ☐ ☐

Comments: _____

_____

60. Board cutter? ☐ ☐ ☐

Comments: _____

_____

61. Wire stitching machine? ☐ ☐ ☐

Comments: _____

_____

	YES	NO	N/A

62. Label-pasting machine? ☐ ☐ ☐

Comments: _____

_____

63. Standing press? ☐ ☐ ☐

Comments: _____

_____

64. Laminating machine? ☐ ☐ ☐

Comments: _____

_____

65. Sign and label makers? ☐ ☐ ☐

Comments: _____

_____

66. Vacuum cleaners? ☐ ☐ ☐

Comments: _____

_____

67. Cleaning supplies and supply carts? ☐ ☐ ☐

Comments: _____

_____

68. Mops, buckets, brooms, and dustpans? ☐ ☐ ☐

Comments: _____

_____

69. Trash compactor? ☐ ☐ ☐

Comments: _____

_____

70. Wastebaskets? ☐ ☐ ☐

Comments: _____

_____

71. Recycling containers? ☐ ☐ ☐

Comments: _____

_____

72. Short and tall ladders and scaffolds? ☐ ☐ ☐

Comments: _____

_____

73. Step stools? ☐ ☐ ☐

Comments: _____

_____

	YES	NO	N/A

74. Moving equipment (dollies, carts)? ☐ ☐ ☐

Comments: _____

_____

75. Storage crates? ☐ ☐ ☐

Comments: _____

_____

K. Behavioral Aspects of Space

1. Has the library's desired atmosphere been considered in planning the building? Atmosphere is the conscious designing of space to create certain effects in users. It represents the interface between the user or customer and the organization. It is a relationship of feelings; it is a rapport with the user. ☐ ☐ ☐

   Comments: _____

   _____

2. Have the following external atmospheric elements been considered:

   a) External signs? ☐ ☐ ☐
   b) Architectural style? ☐ ☐ ☐
   c) Height of the building? ☐ ☐ ☐
   d) Size of the building? ☐ ☐ ☐
   e) Building color? ☐ ☐ ☐
   f) Visibility? ☐ ☐ ☐
   g) Exterior walls? ☐ ☐ ☐
   h) Entrances? ☐ ☐ ☐
   i) Display windows? ☐ ☐ ☐
   j) Views from the building? ☐ ☐ ☐
   k) Building walls? ☐ ☐ ☐
   l) Areas and other structures surrounding the building? ☐ ☐ ☐
   m) Parking availability? ☐ ☐ ☐
   n) Congestion around the building? ☐ ☐ ☐

   Comments: _____

   _____

3. Have the following internal atmospheric elements been considered:

   a) Department locations? ☐ ☐ ☐
   b) Service desk placement? ☐ ☐ ☐
   c) Public address and customer communication systems? ☐ ☐ ☐
   d) Color schemes? ☐ ☐ ☐
   e) Painting and wall surfaces? ☐ ☐ ☐
   f) Composition of ceilings? ☐ ☐ ☐
   g) Music or white noise? ☐ ☐ ☐
   h) Width of aisles? ☐ ☐ ☐
   i) Furniture and fixtures? ☐ ☐ ☐
   j) Movability of furniture and equipment? ☐ ☐ ☐

	YES	NO	N/A
k) Lighting?	☐	☐	☐
l) Scents and sounds?	☐	☐	☐
m) Flooring surfaces (carpeting, tile floors, etc.)?	☐	☐	☐
n) Temperature and humidity?	☐	☐	☐
o) Graphics?	☐	☐	☐
p) Interior displays?	☐	☐	☐
q) Systems for waiting queues?	☐	☐	☐
r) Horizontal transportation?	☐	☐	☐
s) Vertical transportation?	☐	☐	☐

Comments: _____

_____

4. Does arriving at the library create a favorable impression in the minds of customers? Think about how an Apple retail store creates excitement by its design and try to match that in the library building. ☐ ☐ ☐

Comments: _____

_____

5. Do public service staff exhibit a welcoming attitude towards serving library customers? ☐ ☐ ☐

Comments: _____

_____

6. Is there change in the library that will attract users? Window spaces and outdoor displays may encourage use of the library. ☐ ☐ ☐

Comments: _____

_____

7. Has the library space been designed to optimize customer service? Customer service is the experience that library users have in interacting with library staff, collections, and facilities (both electronic as well as physical buildings). ☐ ☐ ☐

Comments: _____

_____

8. In designing the library, have the following factors been considered in establishing how library customer service will be delivered:

a) Have all the physical and electronic points where customers experience service in the library been charted? These are customer "encounter points" or service points and they help create an opinion of the library by its customers. ☐ ☐ ☐

b) Have standards been set for all service points? For example, a standard may be that the time from initial request for materials in an Automated Storage and Retrieval System (ASRS) to delivery at the circulation desk is less than ten minutes. ☐ ☐ ☐

c) Have standards been communicated to library customers? ☐ ☐ ☐

d) Has staff been trained on building and maintaining a good relationship with customers? ☐ ☐ ☐

e) Is ongoing customer service monitored to ensure that standards are met? If you can't measure it, you can't manage it. ☐ ☐ ☐

Comments: _____

_____

9. Has space been designed to meet the unique needs of different segments of customers? Some examples are faculty, students, children, seniors, young adults, and so on. Each segment may have different design needs. ☐ ☐ ☐

    Comments: _____
    _____

10. When designing library spaces, have you considered the following? (Based partially on Joseph Pine's *The Experience Economy*.)

    a) What can be done to improve the aesthetics of the experience for customers? This is what makes your customers want to come into the library and spend time in the library by making the environment more inviting, interesting, and comfortable. ☐ ☐ ☐

    b) What experiences do you want your customers to have in the library, and how will the space be designed to enhance their ability to have them? Is it to take out books, study, attend programs, use the information commons, and so on? Probably a combination of activities and spaces should be designed to accommodate each activity. ☐ ☐ ☐

    c) What do you want your customers to learn from experiencing the library space? How will you encourage them to explore and gain knowledge and experience? ☐ ☐ ☐

    d) How can you make the library experience enjoyable for customers, and encourage them to be repeat customers? What can be done to make the library an entertaining experience? ☐ ☐ ☐

    Comments: _____
    _____

11. Has the following design thinking been used in designing the library space? (Based partially on *Design Thinking* by Stephen J. Ball.)

    a) Do you understand the physical library needs of your target customer population? ☐ ☐ ☐

    b) Do you understand how your customers view your library's products, services and facilities? ☐ ☐ ☐

    c) Have you used marketing research to define what your customers want from their library services? ☐ ☐ ☐

    d) Have you undertaken "anthropological research" by observing what your customer base does in the library? ☐ ☐ ☐

    e) Have you used all the information you gathered to help define the physical library needs of your building? ☐ ☐ ☐

    Comments: _____
    _____

12. Have Feng Shui principles been considered in designing the library:

    a) Does the library space meet the three C's of Feng Shui design; clean, clear, and comfortable? ☐ ☐ ☐

    b) Have the basic principles of the five natural elements (earth, water, fire, metal, and wood) been placed around the library and balanced to increase the flow of chi, or energy? The elements can be balanced internally using interior design and externally with landscape design. Objects are elementally determined by their color, shape, and material so that, for example, the reflective nature of glass windows makes them a water element. Color, texture, and shape will often be the factors when determining the element. ☐ ☐ ☐

    c) Is the library site square or rectangular? ☐ ☐ ☐

    d) Does it have a water view? ☐ ☐ ☐

    e) Is the main entrance unblocked and easily visible when approaching the building? ☐ ☐ ☐

    f) Does the building only have one entrance? A good principle no matter what design principles are used. ☐ ☐ ☐

    g) Has a straight-run stairway been avoided close to the main entry? A U-shaped or split stairway at right angles to the front entry is preferable. ☐ ☐ ☐

    h) Have both yin and yang spaces been used? A space that is considered yang has high ceilings, expansive views, bright colors, or intense lighting. A space that is considered yin would have low ceilings, minimal openings, dark colors, or low lighting. Most libraries are yang in their public spaces. ☐ ☐ ☐

    i) Has a Feng Shui consultant been retained? A consultant can help the architect and interior designer make the library space more attractive and can improve staff and customer well-being. ☐ ☐ ☐

    Comments: _____

    _____

13. Does the library want to encourage physical visits to the library? ☐ ☐ ☐

    Comments: _____

    _____

14. With the access to electronic resources available to customers from their homes, what is the library doing to encourage traveling to the library? How can a trip to the library be an experience? ☐ ☐ ☐

    Comments: _____

    _____

10

Entrepreneurial and Collaborative Spaces

A. Makerspaces

According to *ACRL Tech Connect,* a makerspace is a place where people come together to design and build projects. Makerspaces typically provide access to materials, tools, and technologies to allow for hands-on exploration and participatory learning. It is a collective workshop in which anyone who wants to make or design anything can meet, work, share ideas, pick brains, design software, and use prototyping and fabrication mechanics. Makerspaces are occasionally referred to as fablabs, hackerspaces, or tech shops.

1. Has the library considered having a makerspace? ☐ ☐ ☐

   Comments: _____

   _____

2. Does the library have a biohackerspace? A biohackerspace/lab is a community laboratory that is open to the public where people are encouraged to learn about and experiment with biotechnology. Like a makerspace, a biohackerspace provides people with tools that are usually not available at home. Biotechnology labs need specialized equipment for different types of processes and activities. ☐ ☐ ☐

   Comments: _____

   _____

3. Will parameters be established that define the scope of maker activities? This will influence the type of space required and the environmental aspects of the space ☐ ☐ ☐.

   Comments: _____

   _____

	YES	NO	N/A

4. Will the makerspace be a collaborative effort with one or more organizations? ☐ ☐ ☐

Comments: _____

_____

5. Has the academic library considered partnering with other departments on campus such as engineering and business? ☐ ☐ ☐

Comments: _____

_____

6. How will the library identify space if there is no apparent available space in the existing library? The University of Nevada at Reno determined that patron usage study revealed that students and faculty mostly accessed electronic versions of journals, but print journals took up more than half the library's shelf space. Library officials decided to warehouse low-use items in an automated storage and retrieval system, opening up 18,000 square feet for the makerspace. ☐ ☐ ☐.

Comments: _____

_____

7. Will the makerspace be in the library or in another space available to the library? The Allen County (Indiana) Public Library outfitted a 50 by 10-foot trailer; wired it for phone, power, and Internet; filled it with tools; and then parked it in a lot across the street from the library. ☐ ☐ ☐

Comments: _____

_____

8. Does the library provide 3D printers and 3D scanners? 3D printing, also called rapid prototyping or additive manufacturing, is manufacturing gone digital. A digital design file, created in two dimensions on the screen, can be rendered as a 3-dimensional real-world object. 3D printing is often a component of a makerspace. 3D printers build things by depositing material, layer by layer. This is why the process is described as additive manufacturing. ☐ ☐ ☐

Comments: _____

_____

9. Does the library hacking space have the following areas and environments? (Pettis, 2015)

   a) Open floor space for flexibility and handling large objects? ☐ ☐ ☐
   b) Storage? Provision for scrap storage and trash. ☐ ☐ ☐
   c) Work benches? Can be mobile or secured to the floor. ☐ ☐ ☐
   d) Lighting with fluorescent fixtures spaced depending on bulb length, reflector, and lens types? ☐ ☐ ☐
   e) White ceilings and light-colored walls to help scatter light? ☐ ☐ ☐
   f) Is lighting on one circuit and power tools on another circuit? If you trip the circuit breaker with a tool it won't knock out the lighting. ☐ ☐ ☐
   g) Bin storage? Space to store fasteners, wire connectors, small plumbing parts, and miscellaneous hardware. ☐ ☐ ☐

	YES	NO	N/A

h) Shelves? For power tools in existing plastic cases or in plywood containers made in the shop. ☐ ☐ ☐

i) Flooring that has a concrete base? Clear sealer may be used to make it easier to clean spills. ☐ ☐ ☐

j) Rubber mats for workbenches? This improves traction and reduces fatigue and is also more forgiving to dropped tools. ☐ ☐ ☐

k) Miter saws should be located with clear access on left and right. ☐ ☐ ☐

Comments: _____

10. Does the library hacking space have the following equipment? (Pettis, 2015)

a) Miter saw? Crosscuts construction lumber, molding, or wood for furniture. Put a specialized blade on it and it also cuts plastic pipe and copper tubing. ☐ ☐ ☐

b) Table saw? Rips and crosscuts lumber and shapes parts for woodworking joints. ☐ ☐ ☐

c) Drill press? Makes holes quickly and accurately in wood and metal. ☐ ☐ ☐

d) Planer? Controls the thickness of solid lumber. With some finesse and a few tricks, also flattens stock. ☐ ☐ ☐

e) Joiner? Flattens a wood surface that becomes a reference plane or an edge to be glued to another. ☐ ☐ ☐

f) Band saw? Cuts curves and also re-saws (cuts wood into thinner sections wile wasting less wood than thickness planning). ☐ ☐ ☐

Comments: _____

11. Does the library have a bio-hacking core facility? A core facility is one that contains equipment that might be shared by groups of labs. This equipment includes different types of refrigerators for storing temperature-sensitive reagents, media, and cultures; incubators for growing bacteria and cells; autoclaves for sterilizing media and cultures; and other large pieces of equipment. Some of the equipment that might be considered includes:

a) Household refrigerator (combination refrigerator/freezer)? ☐ ☐ ☐

b) –20 degrees centigrade freezer? ☐ ☐ ☐

c) –86 degrees centigrade freezer? ☐ ☐ ☐

d) Carbon dioxide incubator? ☐ ☐ ☐

e) Biosafety cabinet? ☐ ☐ ☐

f) Shaking incubator? ☐ ☐ ☐

g) Autoclave? ☐ ☐ ☐

h) Refrigerated centrifuge? ☐ ☐ ☐

i) Shaking water bath? ☐ ☐ ☐

j) Pipettors (0 µl to 10 µl or 20 µl; 20 µl to 100 µl; and 100 µl to 1,000 µl)? ☐ ☐ ☐

k) Microcentrifuge? ☐ ☐ ☐

Comments: _____

12. Does the bio-hacking lab have an area for media/buffer/reagent/solution preparation? Media and solution preparation are fundamental activities in biotechnology. Cells and bacteria need media for growth. Chemical reactions and enzymatic processes are carried out in solutions, with buffers and other reagents. Some of the equipment that might be considered includes:

    a) Triple beam balance? ☐ ☐ ☐
    b) Analytical balance? ☐ ☐ ☐
    c) pH meter? ☐ ☐ ☐
    d) Hot plate/magnetic stirrer? ☐ ☐ ☐

    Comments: _____

_____

13. Does the bio-hacking lab have a molecular biology lab? Molecular biology labs require equipment for working with DNA and proteins. Some of the equipment that might be considered includes:

    a) Water bath? ☐ ☐ ☐
    b) DNA gel box? ☐ ☐ ☐
    c) DNA blotter? ☐ ☐ ☐
    d) PCR machine (thermal cycler)? ☐ ☐ ☐
    e) DNA sequencer? ☐ ☐ ☐
    f) UV transilluminator? ☐ ☐ ☐
    g) Protein gel box? ☐ ☐ ☐
    h) Protein blotter? ☐ ☐ ☐
    i) Microtiter plate reader? ☐ ☐ ☐

    Comments: _____

_____

14. Does the bio-hacking lab have an upstream processing space? Upstream processing activities are those that take place during the process of growing eucaryotic cells. These cells might serve as a source of protein or DNA for other activities.

    a) Portable autoclave (for "sterilize in place [SIP] processes")? ☐ ☐ ☐
    b) Spinner bottle bioreactor? ☐ ☐ ☐
    c) Process controlled bioreactor? ☐ ☐ ☐
    d) Microscope? ☐ ☐ ☐
    e) UV visible spectrophotometer? ☐ ☐ ☐
    f) Biolyzer? ☐ ☐ ☐

    Comments: _____

_____

15. Does the bio-hacking lab have a downstream process space? Downstream processing activities are those that occur after the cells have been grown. These activities involve harvesting the cells and purifying specific proteins or other materials.

    a) Tangential flow/diafiltration? ☐ ☐ ☐
    b) Liquid chromatography? ☐ ☐ ☐

    Comments: _____

_____

	YES	NO	N/A

16. Does the library have a "tolerance for noise and smells" that may be generated by a makerspace?

Comments: _____

17. Does the space have its own fire suppression system?

Comments: _____

18. Have all the safety aspects of making things in the library been considered?

Comments: _____

19. Does the makerspace have age or skill requirements in order to use certain tools and machinery?

Comments: _____

20. Are makers required to participate in mandatory safety training in order to use certain tools?

Comments: _____

21. Are makers required to wear protective equipment (eye protection, ear protection, hand protection, etc.)?

Comments: _____

22. Will the library provide personal safety equipment such as goggles, earplugs, gloves, and so on to those who don't have their own?

Comments: _____

23. Is the makerspace ventilated to remove possibly dangerous fumes and gases?

Comments: _____

24. Do biohacking/chemical lab and storage areas designated for science projects have extra precautions?

Comments: _____

25. Is garbage created immediately disposed of in the proper waste bins?

Comments: _____

26. Does the staff have the necessary skills to "instruct and guide" makers?

Comments: _____

27. Has the library developed a resource list of experts who are willing to provide their skills to library customers working in the makerspace?

☐ ☐ ☐

Comments: _____

_____

28. Does the library's insurance policy cover the added risks of a makerspace?

☐ ☐ ☐

Comments: _____

_____

B. Co-working in the Library

1. Does the library have a co-working space? This is a space for freelancers, independent contractors, or small, web-based businesses—people who "work from home." They appreciate the necessities of an actual office while in a public space: comfy seating, space to spread out documents, outlets always nearby, updated scanner/copier/fax machine, the ability to use cell phones, and maybe even a chance to collaborate.

☐ ☐ ☐

Comments: _____

_____

2. Does access to the space require residence in the library's community for public libraries or affiliation with the educational institution for university libraries?

☐ ☐ ☐

Comments: _____

_____

3. Is the space free or is there a fee membership to have access to the business space?

☐ ☐ ☐

Comments: _____

_____

4. Does the library require training or an orientation session before the user is able to access equipment and facilities in the space?

☐ ☐ ☐

Comments: _____

_____

5. Does the library have computers available to the public loaded with office-type software (Microsoft Word, Excel, PowerPoint, Adobe Photoshop, etc.)?

☐ ☐ ☐

Comments: _____

_____

6. Does the space allow for studio-type activities?

   a) Digital production lab? ☐ ☐ ☐
   b) Voice-over studio? ☐ ☐ ☐
   c) Main production studio? ☐ ☐ ☐

Comments: _____

_____

	YES	NO	N/A

7. Does the library have conference rooms available to the public? ☐ ☐ ☐

   Comments: _____

   _____

8. May conference rooms be scheduled in advance and if so, what is the time horizon for reservations? ☐ ☐ ☐

   Comments: _____

   _____

9. Are there restrictions on the activities that may take place in the conference rooms? ☐ ☐ ☐

   Comments: _____

   _____

10. Is an Espresso Book Machine available? The patented Espresso Book Machine (EBM) makes a paperback book in minutes, at point of need. The technology is excellent for self- or custom publishing. ☐ ☐ ☐

    Comments: _____

    _____

11. Does the space allow areas for coffee and refreshments as well as an environment for collaboration? ☐ ☐ ☐

    Comments: _____

    _____

12. Has the space been designed so that it is accessible to customers when the library is closed? ☐ ☐ ☐

    Comments: _____

    _____

C. Technology Lending Library

1. Does the library have a technology lending library? This may include an array of standard and experimental technologies for students and faculty to develop course-related and personal projects. ☐ ☐ ☐

   Comments: _____

   _____

2. Does the library have secure and sufficient storage to house and protect the equipment for loan? ☐ ☐ ☐

   Comments: _____

   _____

3. Does the library have policies in place to govern the loaning of technology equipment? ☐ ☐ ☐

   Comments: _____

   _____

	YES	NO	N/A

4. Is the equipment bar coded or identified with an RFID tag? ☐ ☐ ☐

Comments: _____

_____

5. Is the equipment listed in the library's catalog? ☐ ☐ ☐

Comments: _____

_____

6. Is the following equipment available for loan?

a) Digital cameras and camcorders? Devices for capturing video and still images. ☐ ☐ ☐

b) Laptops and netbooks? Laptops and netbooks with either Mac, Windows, or Chrome operating systems. Some libraries offer iWatch checkout. ☐ ☐ ☐

c) Tablets and iPads? Apple and Android tablets. ☐ ☐ ☐

d) Cables and accessories? Cables and small peripherals. ☐ ☐ ☐

e) Makerspace components? Devices for tinkering and prototyping. ☐ ☐ ☐

f) Gaming devices? Game consoles, controllers, and games for checkout or for library use. ☐ ☐ ☐

g) DSLR cameras and accessories? DSLR, or digital single-lens reflex, cameras are higher-end than the point-and-shoot variety and allow users to take higher-quality photos. ☐ ☐ ☐

h) Digital media equipment? For use in the library's digital media lab, music rooms, and media production studios. ☐ ☐ ☐

i) Projectors and media players? This category includes projectors, slide projectors, VCRs, and DVD players. ☐ ☐ ☐

j) Calculators? Scientific, financial and graphing calculators.

k) E-readers? Devices used to display digital books. ☐ ☐ ☐

l) Design and modeling tools? Includes calipers, drawing forms, graphing tablets, and modeling kits. ☐ ☐ ☐

Comments: _____

_____

D. Music Instrument Lending Library

1. Does the library provide musical instruments for loan? ☐ ☐ ☐

Comments: _____

_____

2. Does the library employ special storage containers for the instruments? ☐ ☐ ☐

Comments: _____

_____

3. Do the instruments have bar codes or RFID tags? ☐ ☐ ☐

Comments: _____

_____

	YES	NO	N/A

4. In university libraries, are the instruments available to students and faculty? ☐ ☐ ☐

   Comments: _____

   _____

5. In public libraries, are the instruments available to adults and children? ☐ ☐ ☐

   Comments: _____

   _____

6. Are the instruments used in the library or may they be taken out on loan? ☐ ☐ ☐

   Comments: _____

   _____

7. What types of instruments are available?

   a) Banjos ? ☐ ☐ ☐
   b) Bongos? ☐ ☐ ☐
   c) Buffalo drum? ☐ ☐ ☐
   d) Djembe drum? ☐ ☐ ☐
   e) Electronic keyboard? ☐ ☐ ☐
   f) Guitar? ☐ ☐ ☐
   g) Kalimba? ☐ ☐ ☐
   h) Mountain dulcimer? ☐ ☐ ☐
   i) Ukulele? ☐ ☐ ☐
   j) Violin? ☐ ☐ ☐

   Comments: _____

   _____

E. Tool Lending Library

1. Is the loan of tools free or is there a fee? ☐ ☐ ☐

   Comments: _____

   _____

2. Are the tools located in the library or in adjacent sheds or outbuildings? ☐ ☐ ☐

   Comments: _____

   _____

3. Are the tools listed in the library's catalog? ☐ ☐ ☐

   Comments: _____

   _____

4. Do the tools have bar code or RFID identification labels? ☐ ☐ ☐

   Comments: _____

   _____

	YES	NO	N/A

5. Are the following types of tools available?

    a) Weed eaters? ☐ ☐ ☐

    b) Extension cords? ☐ ☐ ☐

    c) Hedge trimmers? ☐ ☐ ☐

    d) Demolition hammers? ☐ ☐ ☐

    e) Electric snakes? ☐ ☐ ☐

    f) Ladders? ☐ ☐ ☐

    g) Cement mixers? ☐ ☐ ☐

    h) Hand trucks and dollies? ☐ ☐ ☐

    i) Plumbing tools? ☐ ☐ ☐

    j) Pipe threaders? ☐ ☐ ☐

    k) Circular saws? ☐ ☐ ☐

    l) Sanders? ☐ ☐ ☐

    m) Table saws? ☐ ☐ ☐

    n) Drills? ☐ ☐ ☐

    o) Wheelbarrows? ☐ ☐ ☐

    p) Post hole diggers? ☐ ☐ ☐

    Comments: _____

    _____

6. Does the tool library carry gas-powered or power-actuated tools? If so, does the library have release forms holding the library harmless when customers borrow the tools? ☐ ☐ ☐

    Comments: _____

    _____

7. Does the library provide a collection of how-to books, videos, and DVDs that may be borrowed? ☐ ☐ ☐

    Comments: _____

    _____

8. Does the library employ tool-lending specialists to advise customers about what tools are required and how to use them? ☐ ☐ ☐

    Comments: _____

    _____

9. Does the library collaborate with local hardware and big-box stores to provide workshops and assistance? ☐ ☐ ☐

    Comments: _____

    _____

F. Seed Lending Library

1. Does the library have a seed library? Gardeners "check out" seeds from the library, plant them in their own gardens, and harvest seeds from the plants they grew and return the seeds to the library. In some communities this may be illegal. ☐ ☐ ☐

   Comments: _____

   _____

2. Is seed sharing legal in the community and/or state? ☐ ☐ ☐

   Comments: _____

   _____

3. Does the library need a permit to share seeds? ☐ ☐ ☐

   Comments: _____

   _____

4. How does the library acquire seeds?

   a) By soliciting patrons, local volunteer groups, social groups, churches, and so on for seed packets meeting their specifications (traditional, heirloom, organic, etc.)? ☐ ☐ ☐

   b) By using tax revenues and/or monetary donations to purchase seeds from local retail stores in the community? ☐ ☐ ☐

   c) By accepting donated seeds from seed distributors and retailers? ☐ ☐ ☐

   Comments: _____

   _____

5. How does the library "store" the seeds for loan? Old card catalogs are sometimes used. ☐ ☐ ☐

   Comments: _____

   _____

6. Are seed packets cataloged and included in the library's database? ☐ ☐ ☐

   Comments: _____

   _____

Materials Handling and Storage: Book Stacks and Shelving

A. Conventional Stationary Stacks and Shelving

1. Has sufficient shelving been planned to meet the current and future needs of the library? (Consider the size of the current collection, growth of the collection for at least 20 years, and the percentage of the collection that will be out on loan. A rule of thumb for roughly calculating shelving requirements is to assume 150 volumes per 3-foot single-faced section, 7 shelves high.)

☐ ☐ ☐

Comments: _____

_____

2. With the transition to electronic resources, have future plans estimated the loss of storage space required for print/physical collections?

☐ ☐ ☐

Comments: _____

_____

3. If print collections shrink, and electronic collections grow, has the number of electronic reading devices required been calculated?

☐ ☐ ☐

Comments: _____

_____

4. Is the shelving selection based on the standard titled American National Standard for Single-Tier Steel Bracket Library Shelving, ANSI/NISO Z39.73–1994? This is the standard for shelving.

☐ ☐ ☐

Comments: _____

_____

5. Are book stacks arranged sequentially in parallel ranges so that users can easily locate materials? ☐ ☐ ☐

    Comments: _____

    _____

6. If book stacks are not arranged sequentially in parallel ranges, are variations clearly indicated? ☐ ☐ ☐

    Comments: _____

    _____

7. Are there labels on both ends of ranges of book stacks? ☐ ☐ ☐

    Comments: _____

    _____

8. Is display shelving included to merchandise the collection? For example, are there:

    a) Display units with sloping shelves? ☐ ☐ ☐
    b) Point-of-purchase displays as seen in bookstores and department stores? ☐ ☐ ☐
    c) Spinners or towers? ☐ ☐ ☐
    d) Slatwall end panels or wall units? ☐ ☐ ☐

    Comments: _____

    _____

9. Are all stacks and shelves clearly labeled as to content on both end panels and shelf lips? ☐ ☐ ☐

    Comments: _____

    _____

10. Are there attempts to break the monotony of shelving by creative arrangement of seating or height and/or type of shelving? ☐ ☐ ☐

    Comments: _____

    _____

11. Are there no more than eight 36-inch shelving sections without a break? ☐ ☐ ☐

    Comments: _____

    _____

12. Have length of shelving and width of aisles been determined on the basis of traffic patterns and user accessibility? (See ADA guidelines in chapter 7.) ☐ ☐ ☐

    Comments: _____

    _____

13. Is the shelving height and depth adequate for users? ☐ ☐ ☐

    Comments: _____

    _____

14. Do double-faced sections of bracket shelving have bases 20 or 40 inches deep? ☐ ☐ ☐

    Comments: _____

    _____

15. Is freestanding shelving, ranging from seventy-eight inches or higher, anchored to the floor, or braced with top tie struts? ☐ ☐ ☐

Comments: _____
_____

16. Are the shelving units:

   a) Sturdy and well built? ☐ ☐ ☐
   b) Able to bear prescribed loads without sagging, bending, leaning, swaying, or collapsing? ☐ ☐ ☐
   c) Equipped with a finish that will endure normal use and cleaning for at least thirty years without signs of wear? ☐ ☐ ☐
   d) Smoothly finished with no burrs or sharp edges? ☐ ☐ ☐
   e) Standardized in design and color? ☐ ☐ ☐
   f) Designed to have interchangeable parts? ☐ ☐ ☐
   g) Equipped with adjustable shelves? ☐ ☐ ☐
   h) Equipped with shelves that are relatively easy to move when they are unloaded? ☐ ☐ ☐
   i) Equipped with shelves that are relatively easy to move when they are loaded? ☐ ☐ ☐
   j) Braced and/or anchored to comply with local regulations? ☐ ☐ ☐
   k) Equipped with end panels? ☐ ☐ ☐
   l) Equipped with canopies? ☐ ☐ ☐

Comments: _____
_____

17. Are there special features such as:

   a) Pullout shelves? ☐ ☐ ☐
   b) Built-in lighting? ☐ ☐ ☐
   c) Electrical access? ☐ ☐ ☐
   d) Shelf dividers? ☐ ☐ ☐
   e) Movable book supports of adequate size? ☐ ☐ ☐
   f) Range-label holders? ☐ ☐ ☐
   g) Shelf-label holders? ☐ ☐ ☐
   h) Current periodical shelves? ☐ ☐ ☐
   i) Wide-lip newspaper shelves? ☐ ☐ ☐
   j) Atlas and dictionary stands? ☐ ☐ ☐

Comments: _____
_____

18. Are there accessories to display and house:

   a) CDs? ☐ ☐ ☐
   b) DVDs? ☐ ☐ ☐
   c) Audiocassettes? ☐ ☐ ☐
   d) Videocassettes? ☐ ☐ ☐
   e) Picture books? ☐ ☐ ☐
   f) Paperback books? ☐ ☐ ☐
   g) Oversized and miniature materials? ☐ ☐ ☐
   h) Archival materials? ☐ ☐ ☐

i)   Films, filmstrips, slides, microforms? ☐ ☐ ☐

j)   Realia? ☐ ☐ ☐

k)   Other odd-shaped items? ☐ ☐ ☐

Comments: _____

_____

19. Does periodical shelving have a maximum reach height of forty-eight inches? ☐ ☐ ☐

Comments: _____

_____

20. Does periodical shelving have sloping shelves that tilt and allow for storage on a flat shelf beneath? ☐ ☐ ☐

Comments: _____

_____

21. Is the edge or lip on newspaper shelves wide enough to hold a large Sunday edition? (Use the Sunday *New York Times* as a test.) ☐ ☐ ☐

Comments: _____

_____

22. Is there a need for enclosed shelving with lockable doors? ☐ ☐ ☐

Comments: _____

_____

23. Are there shelf/table units for reference and index materials? ☐ ☐ ☐

Comments _____

_____ :

24. Have nonpublic work and storage areas been provided with appropriate shelving? ☐ ☐ ☐

Comments: _____

_____

25. Have the following aisle allowances been made?

a)   Are cross aisles that run parallel to the stack aisles and that are intended to break up the side aisles into increments (also known as range aisles) a minimum of 36 inches wide to meet accessibility requirements? Cross aisles are required to be 36 inches wide to meet accessibility requirements. Some libraries prefer 42 to 48 inches. ☐ ☐ ☐

b)   Are end aisles that run perpendicular to side aisles and may have books on one side a minimum of 36 inches wide? Occasionally 44 inches is required in some cases. ☐ ☐ ☐

c)   Are main aisles a minimum of forty-four inches wide? These aisles serve as the principle access routes. ☐ ☐ ☐

Comments: _____

_____

B. Movable-Aisle Compact Shelving

1. Is there a need for movable-aisle compact shelving? This type of shelving is sometimes referred to as compact shelving. These are shelving systems that ride on moveable carriages over floor-installed rails. The benefit of this shelving is to maximize use of floor space by having one access aisle, which is located by moving the carriage-mounted shelving in a manner that opens an access aisle at a desired location. Mobile shelving sits on carriages on top of tracks that can be moved to open up stacks for access to books. ☐ ☐ ☐

    Comments: _____

    _____

2. If the rails cannot be recessed, will there be some kind of deck for the system? ☐ ☐ ☐

    Comments: _____

    _____

3. Is the building capable of holding the substantial weight of a compact installation? (Generally, floor load capacity for compact shelving is 300 pounds live load per square foot, or double the amount required for conventional shelving.) ☐ ☐ ☐

    Comments: _____

    _____

4. Does the cost of the space saved justify the cost of the system? ☐ ☐ ☐

    Comments: _____

    _____

5. Are all ADA and safety codes met? ☐ ☐ ☐

    Comments: _____

    _____

6. What type of system is being considered?
    a) A manual system operates with a push or pull of the handle to move the shelving units stored on carriages that move from side to side. Shelving systems can go from two to four sections of shelving deep. Sections of shelving are easily moved to the side, exposing another section of shelving behind the original layer. Human power is strained after the number of sections exceeds four. ☐ ☐ ☐
    b) A mechanical assistance system moves the stacks with a turn of a handle attached to a pulley or a lever at the end of the stack range that is connected to a gear system that drives the range. Mechanical assist mobile shelving systems require only the turn of a handle to access shelves. Usually one pound of pressure can move thousands of pounds of steel and media. ☐ ☐ ☐
    c) An electrical system moves the stacks by pushing a button on a control panel activating one or more electrical motors. Some systems may be configured with multiple aisles without compromising safety. There usually is a system indicator light that alerts users of the unit's operational status. The control panel may be located to meet ADA requirements. ☐ ☐ ☐

    Comments: _____

    _____

	YES	NO	N/A

7. If an electrical system is used, does it have a manual override? ☐ ☐ ☐

 Comments: _____

 _____

8. Does the electrical system have:

 a) Emergency stop button? ☐ ☐ ☐
 b) Automatic lock after displacement? ☐ ☐ ☐
 c) Protection against radio frequency interferences? ☐ ☐ ☐
 d) Toe-level infrared beam to detect people in the stacks? ☐ ☐ ☐

 Comments: _____

 _____

9. Is there a "fail-safe system" that stops the movement of the units if an obstacle (especially a person) is encountered? ☐ ☐ ☐

 Comments: _____

 _____

10. Have the specialized cleaning and maintenance needs of compact shelving been considered? ☐ ☐ ☐

 Comments: _____

 _____

11. Under the purchase agreement, how long will the vendor maintain the system? ☐ ☐ ☐

 Comments: _____

 _____

12. Will movable stacks be accessible to the public with or without staff assistance? ☐ ☐ ☐

 Comments: _____

 _____

13. If the movable stacks are not accessible to the public without staff assistance, what sort of devices will be employed to keep the movable stacks from operating? ☐ ☐ ☐

 Comments: _____

 _____

14. Can the system be expanded? ☐ ☐ ☐

 Comments: _____

 _____

15. Can the system be moved easily to another location? ☐ ☐ ☐

 Comments: _____

 _____

16. Does the library have mobile shelving? Mobile shelving moves out of the way to make room for special events, and easily moves back in place. ☐ ☐ ☐

 Comments: _____

 _____

C. Automated Storage and Retrieval Systems (ASRS)

1. Is a mechanized book-retrieval system needed? In order to determine if an ASRS is justified, a cost-benefit analysis should be done measuring the floor space that will be gained (or not built) against the cost of the system and the potential inconvenience to users. An ASRS typically works as follows: ☐ ☐ ☐

   a) Books and media are stored in bins on a rack structure. The bins are arranged by aisles, each of which has a "mini-load crane" guided by rails at top and bottom. ☐ ☐ ☐
   b) Books and media are stored in bins in the ASRS by bar code or RFID tag. ☐ ☐ ☐
   c) Requests for retrieval of ASRS items are submitted via the library catalog, and are transmitted electronically to a designated circulation desk. ☐ ☐ ☐
   d) The request triggers the ASRS automatic crane in the appropriate aisle to deliver the bin to a designated pickup station. ☐ ☐ ☐

   Comments: _____
   _____

2. Have the following been considered before deciding upon a mechanized book-retrieval system:

   a) Size of the collection? ☐ ☐ ☐
   b) Space available for the collection? ☐ ☐ ☐
   c) Annual growth of the collection? ☐ ☐ ☐
   d) Policy as to what will be stored in the ASRS? ☐ ☐ ☐
   e) Policy on how long customers will have to wait for a book after it is requested from the ASRS? ☐ ☐ ☐
   f) Cost of staffing the ASRS and the circulation desk? ☐ ☐ ☐
   g) Cost of installation and maintenance? ☐ ☐ ☐
   h) Load-bearing capabilities of the building? ☐ ☐ ☐
   i) Contingency plans if the ASRS fails? ☐ ☐ ☐

   Comments: _____
   _____

3. Will installation of the system require any organizational changes that may impact the physical library building? ☐ ☐ ☐

   Comments: _____
   _____

D. Materials Handling Systems

1. Has a materials handling system been considered? This is a self-service return and sorting system that automates material handling tasks such as sorting and returning materials to the shelf. These systems employ RFID tags for material handling. ☐ ☐ ☐

   Comments: _____
   _____

2. Does the system provide the following services:

   a) Automated check-in and sorting? When a book, CD, or other item is returned to the system, it drops onto an automated conveyor system that carries materials to a sorting area. The book is actually checked back in to the library's circulation system by radio frequency identification (RFID) technology—an RFID antenna detects an RFID chip on the returned book. ☐ ☐ ☐

   b) Automated routing and sorting within the library? If the materials will stay at the library where they are returned, they are routed to sorting machines that place items on book carts for reshelving. ☐ ☐ ☐

   c) Automated routing and sorting to another library? If the materials are directed to another library, they are routed to a series of motorized bins for delivery to another location. These bins have movable floors that raise or lower as items are added or removed. This feature prevents staff from having to bend and reach deep into the bin to remove items from the bottom. ☐ ☐ ☐

   Comments: _____
   _____

3. Do the elements of the system consist of:

   a) Automated check-in? ☐ ☐ ☐
   b) Automated checkout? ☐ ☐ ☐
   c) Reserve pickup? ☐ ☐ ☐
   d) Circulation materials sorting? ☐ ☐ ☐
   e) Patron reserve retrieval and storage systems? ☐ ☐ ☐
   f) Conveyor lift units? ☐ ☐ ☐
   g) RFID tracking? ☐ ☐ ☐
   h) Circulation materials conveying systems? ☐ ☐ ☐

   Comments: _____
   _____

4. Does the system provide the following benefits:

   a) Reduce the time it takes to get books and other materials back in circulation after they've been returned? ☐ ☐ ☐

   b) Reduce the risk of personal injury to library staff due to the repetitive motion and physical strain involved in sorting and moving books on a regular basis? ☐ ☐ ☐

   c) Allow self-checkout? This is usually faster than interacting with staff, especially if a number of materials are being checked out at the same time. ☐ ☐ ☐

   d) Provide greater security and inventory control through RFID? This technology usually enhances materials security, reducing theft and associated costs. ☐ ☐ ☐

   Comments: _____
   _____

E. Remote Storage

1. Has an off-site storage facility been considered as a place to house secondary or little-used materials?

    $\square$ $\square$ $\square$

    Comments: _____

    _____

2. Has a policy been established determining what will be placed in remote storage and what will remain in the library?

    $\square$ $\square$ $\square$

    Comments: _____

    _____

3. Will the remote storage facility serve one library or will it store books cooperatively from a number of libraries?

    $\square$ $\square$ $\square$

    Comments: _____

    _____

4. If the facility will serve as a cooperative storage facility for a number of libraries, have policies been negotiated to meet the needs of all members of the cooperative?

    $\square$ $\square$ $\square$

    Comments: _____

    _____

5. Is the facility a cold-storage warehouse that maximizes the use of space through high-density shelving?

    $\square$ $\square$ $\square$

    Comments: _____

    _____

6. Has a policy decision been made to store materials by size or by subject classifications?

    $\square$ $\square$ $\square$

    Comments: _____

    _____

7. Does the remote storage facility have a high degree of book storage density, narrow aisles, long ranges of shelving, and shelving by size divisions in order to approach maximum density? Size vs. subject is usually the standard policy.

    $\square$ $\square$ $\square$

    Comments: _____

    _____

8. Will compact shelving units be used?

    $\square$ $\square$ $\square$

    Comments: _____

    _____

9. Is the off-site storage facility designed to environmentally protect materials?

    a) Is the facility insulated and does it have a high degree of air tightness and vapor barriers as compared to standard construction? Usually this means twice the insulation value, six times the air tightness, and the vapor barrier fifteen times that of standard construction.

    $\square$ $\square$ $\square$

b) Does the HVAC system provide an environment of approximately 55 to 65 degrees F and 40 to 55 percent relative humidity regardless of seasonal changes? ☐ ☐ ☐

c) Is the HVAC designed to quickly react to changes in the exterior climate? ☐ ☐ ☐

Comments: _____

_____

10. Is low lighting (sodium vapor or fluorescent light fixtures with UV shields) used in order to reduce the damage that light does to books? ☐ ☐ ☐

Comments _____

_____:

11. Is the amount of time that lights are left on kept to a minimum? ☐ ☐ ☐

Comments: _____

_____

12. Is public access available at the site for faculty, students, library users, and so on? ☐ ☐ ☐

Comments: _____

_____

13. If access is available, may customers browse the facility or only read? ☐ ☐ ☐

Comments: _____

_____

14. Is there space for processing in the facility? ☐ ☐ ☐

Comments: _____

_____

15. Is there a high-security area for special collections? ☐ ☐ ☐

Comments: _____

_____

16. Is the location of stored materials linked through bar codes, RFID tags, or inventory control numbers to the library's catalog? ☐ ☐ ☐

Comments: _____

_____

17. If stored materials are not linked to the library's catalog, is there an inventory control system used at the storage facility to easily locate and retrieve materials? ☐ ☐ ☐

Comments: _____

_____

18. Has a delivery system been established for quick access to the stored collections for library customers? ☐ ☐ ☐

Comments: _____

_____

19. Have customer service standards been established for time required to retrieve materials from remote storage back to the library?

☐ ☐ ☐

Comments: _____

_____

20. Has an electronic document system been established to enhance on-site access of remote collections and reduce the number of materials that will need to be stored off-site?

☐ ☐ ☐

Comments: _____

_____

12

Building Systems

A. Acoustics

1. Is an acoustic engineer part of the architect's design team?

    Comments: _____
    _____

2. Has the ambient noise level been considered in locating the library building?

    Comments: _____
    _____

3. If the selected site is noisy, has location of the building been sited to mitigate ambient noise as much as possible?

    Comments: _____
    _____

4. How does the library deal with noise?

    a) Does the library want noise control? With noise control the library typically looks to block noise through noise barriers masking the noise from equipment, facilities, and the surroundings.

    b) Does the library want soundproofing? With soundproofing the library is typically looking to absorb sound or deal with echo and reverberation problems.

    Comments: _____
    _____

5. Is the library too quiet? If it is too quiet the slightest sounds may cause distractions for users. ☐ ☐ ☐

   Comments: _____

   _____

6. Are circulation, information, and reference service points located and designed so noise will not disrupt other areas? ☐ ☐ ☐

   Comments: _____

   _____

7. Have restrooms, conference rooms, lounges, photocopiers, and public telephones been located where the noise will be the least distracting? ☐ ☐ ☐

   Comments: _____

   _____

8. Have the following elements been chosen to contribute to noise reduction:

   a) Carpeting? ☐ ☐ ☐
   b) Other floor surfaces that do not generate and/or transfer noise? ☐ ☐ ☐
   c) Wall coverings? ☐ ☐ ☐
   d) Window coverings? ☐ ☐ ☐
   e) Ceiling surfaces? ☐ ☐ ☐
   f) Furniture? ☐ ☐ ☐
   g) Shelving? ☐ ☐ ☐
   h) Equipment? ☐ ☐ ☐
   i) Ceiling heights? ☐ ☐ ☐

   Comments: _____

   _____

9. Have quiet-opening doors and windows been selected? ☐ ☐ ☐

   Comments: _____

   _____

10. Has equipment in public areas (computer printers, photocopiers, etc.) been chosen for quiet operation? ☐ ☐ ☐

    Comments: _____

    _____

11. Has a sound masking system been considered? A sound masking system renders noise and conversation more difficult—or impossible—to comprehend. It also reduces dynamic range (the variation in sound over time), making a space feel quieter. ☐ ☐ ☐

    Comments: _____

    _____

12. Is there background sound, such as the ventilating system or other "white noise" sources, to mask minor distracting sounds? ☐ ☐ ☐

    Comments: _____

    _____

13. What sound levels are desired for the library? Sound levels are measured in units of decibels (dB). Suggested sound levels are:

    a) Quiet reading rooms; 30dB?   ☐ ☐ ☐

    b) Open reading areas; 45dB?   ☐ ☐ ☐

    c) Areas where normal conversation takes place; 60–70dB?   ☐ ☐ ☐

    d) Circulation desks, workrooms; 60–75dB?   ☐ ☐ ☐

Comments: _____

_____

14. If the exterior environment of the library has a high noise level, has the location and type of window been considered in planning fenestration?   ☐ ☐ ☐

Comments: _____

_____

15. Are traffic patterns throughout the building designed to keep noise and confusion away from readers?   ☐ ☐ ☐

Comments: _____

_____

16. Are there acoustically controlled quiet areas and are they accessible from widely distributed entrance points?   ☐ ☐ ☐

Comments: _____

_____

17. Are soundproof rooms available?   ☐ ☐ ☐

Comments: _____

_____

18. Are there areas where furniture is arranged to discourage conversation?   ☐ ☐ ☐

Comments: _____

_____

19. In order to control noise levels in public rooms, have:

    a) Walls been extended from floor to the structural deck above?   ☐ ☐ ☐

    b) Ceiling tiles been selected that have a good sound insulation quality?   ☐ ☐ ☐

    c) Air ducts been insulated and balanced to reduce air noise?   ☐ ☐ ☐

    d) Mechanical equipment rooms not located close to public or staff areas?   ☐ ☐ ☐

Comments: _____

_____

20. Is there acoustical separation between public and staff areas?   ☐ ☐ ☐

Comments: _____

_____

21. Have suspended ceilings with noise-absorbing materials been considered?   ☐ ☐ ☐

Comments: _____

_____

22. Have HVAC systems been specified to have an ambient sound level compatible with the occupancy? If the HVAC system is too noisy, conversation may be difficult. If the HVAC is too quiet, unwanted conversations and other distracting noises will be heard. Usually quieter systems cost more but in the long run are worth it.

☐ ☐ ☐

Comments: _____

_____

23. Has noise from the HVAC been controlled? Heating, ventilating, and air-conditioning ductwork can be a source of noise as well as a transmitter of it.

☐ ☐ ☐

Comments: _____

_____

24. Does the HVAC ductwork have insulation to mitigate noise? Sheet metal ductwork without insulation can produce popping and banging noises due to expansion and contraction caused by changes in air temperature. Components within the duct system, abrupt changes in direction, and restrictions in the system can produce turbulence and air rush noise.

☐ ☐ ☐

Comments: _____

_____

25. Does the HVAC ductwork have:

a) Fiberglass duct liner? Designed for installation inside sheet metal ductwork to attenuate air rush and central equipment noise as well as to control heat loss or gain through duct walls.

☐ ☐ ☐

b) Fiberglass duct board? This product combines acoustical/thermal insulation with a reinforced foil-scrim-kraft air barrier/vapor retarder, from which complete air duct systems may be fabricated.

☐ ☐ ☐

Comments: _____

_____

26. Are mechanical systems (elevators, heating, and air conditioning equipment) located away from quiet areas and/or acoustically shielded?

☐ ☐ ☐

Comments: _____

_____

27. Have noisy equipment and activities been located where the sound may be dampened, or have they been located in remote areas?

☐ ☐ ☐

Comments: _____

_____

28. Have unusual architectural features of the building (domes, atriums, towers, etc.) been studied to determine that they will not create whistling or other atmospheric noise conditions inside the building?

☐ ☐ ☐

Comments: _____

_____

B. HVAC (Heating, Ventilation, and Air Conditioning) Systems

1. Have all of the main functions of an HVAC system been considered?

   a) Heating which maintains adequate room temperature, especially during colder weather conditions. ☐ ☐ ☐

   b) Ventilation, which is associated with air movement. Adequate ventilation allows carbon dioxide to go out and oxygen to get in, allowing library users to inhale fresh air. ☐ ☐ ☐

   c) Air-conditioning system that controls temperature as well as ventilation. ☐ ☐ ☐

   d) Humidity control that is important in libraries, especially because of the paper and other media housed in the building. This includes both humidification (adding moisture to the building) and dehumidification (reducing moisture in the building). ☐ ☐ ☐

   Comments: _____

   _____

2. Is the HVAC system:

   a) Simple to operate? ☐ ☐ ☐

   b) Easy to maintain? ☐ ☐ ☐

   c) Efficient to run? ☐ ☐ ☐

   Comments: _____

   _____

3. Does the system, including ductwork, make efficient use of space? ☐ ☐ ☐

   Comments: _____

   _____

4. Can temperature and humidity be zone controlled room by room, either centrally or from lockable thermostats? ☐ ☐ ☐

   Comments: _____

   _____

5. Is there a remote system of control such as BACnet? ☐ ☐ ☐

   a) BACnet is a data communications protocol for Building Automation and Control networks. It was developed under the auspices of the American Society of Heating, Refrigerating and Air-Conditioning Engineers (ASHRAE). BACnet is an American national standard, a European standard, and a national standard in more than thirty countries, as well as an International Organization for Standardization global standard.

   b) A BACnet system can control:

      i) HVAC systems

      ii) Fire detection and alarm

      iii) Lighting control

      iv) Security

      v) Smart elevators

      vi) Utility company interface

   c) The system allows monitoring of building systems as well as adjustments of the systems from remote locations.

   Comments: _____

   _____

6. Is the building properly insulated to help maintain temperature efficiently? ☐ ☐ ☐

   Comments: _____

   _____

7. If the building has large windows or skylights, is there provision for maintaining temperature through window coverings or special glazing? ☐ ☐ ☐

   Comments: _____

   _____

8. Is there adequate ventilation using:

   a) A mechanical air-exchange system? ☐ ☐ ☐
   b) Natural ventilation? ☐ ☐ ☐

   Comments: _____

   _____

9. Is there provision for ventilation if the climate control fails? ☐ ☐ ☐

   Comments: _____

   _____

10. Do the windows open, and if so is there control by staff? ☐ ☐ ☐

    Comments: _____

    _____

11. Do open windows present the opportunity for book and media theft? ☐ ☐ ☐

    Comments: _____

    _____

12. Can environmental pollution be filtered out of the air? ☐ ☐ ☐

    Comments: _____

    _____

13. Can relative humidity (RH) be controlled for various types of media within a 5 percent variance:

    a) Books and Paper: 40–55 percent RH? ☐ ☐ ☐
    b) Books, Papers, and Photos: 50 percent RH? ☐ ☐ ☐
    c) Magnetic Media: 30 percent RH? ☐ ☐ ☐

    Comments: _____

    _____

14. Are temperature/humidity conditions appropriate for:

    a) Rare materials? ☐ ☐ ☐
    b) Special collections? ☐ ☐ ☐
    c) Archives? ☐ ☐ ☐
    d) Electronic workstations, information commons, and telecommunication rooms? ☐ ☐ ☐
    e) Public areas? ☐ ☐ ☐
    f) Staff work areas? ☐ ☐ ☐
    g) Closed stacks? ☐ ☐ ☐

    Comments: _____

    _____

15. Are there emergency backup generators that can be used if the electricity goes off and the air conditioning shuts down? ☐ ☐ ☐

    Comments: _____

    _____

16. Has the HVAC been properly designed and installed? Proper heating, ventilating, and air-conditioning are key to maintaining a comfortable, healthy, and productive work environment. Collectively, these systems account for approximately 40 percent of the electricity used in commercial buildings. Improved heating and cooling performance, along with substantial energy savings, can be achieved by implementing energy-efficiency measures. ☐ ☐ ☐

    Comments: _____

    _____

17. Have the cooling and heating load requirements been calculated? Accurate load requirement calculations (how much heating and cooling you actually need) will provide a system that meets the HVAC needs without adding excess capacity. Over-sizing equipment increases the capital cost at the time of the installation and the costs of operation of the equipment. ☐ ☐ ☐

    Comments: _____

    _____

18. Have possible ways to reduce the cooling and heating loads been considered?

    a) Has reducing solar gain been considered? Solar gain should be reduced in warm climates and increased in cold climates. Solar gain may be reduced by using a cool roof and window tints. ☐ ☐ ☐

    b) Have the most efficient office equipment and consumer products been specified to reduce heat gain? ☐ ☐ ☐

    Comments: _____

    _____

19. Have energy recovery ventilation systems been considered to reclaim waste energy? Using recovered energy from the exhaust air helps condition the incoming fresh air. ☐ ☐ ☐

    Comments: _____

    _____

20. Has supplemental dehumidification been considered in humid climates? By controlling humidity at your facility, you can increase occupant comfort and allow for further downsizing of equipment. ☐ ☐ ☐

    Comments: _____

    _____

21. In dry climates, have evaporative coolers been considered? These coolers use the evaporation of water to cool spaces, eliminating the need for energy-intensive compressors. ☐ ☐ ☐

    Comments: _____

    _____

22. Are programmable thermostats used to achieve peak energy savings? They may be controlled to reduce requirements when the building is closed.

☐ ☐ ☐

Comments: _____

_____

23. Are the thermostats only programmable by authorized library staff?

☐ ☐ ☐

Comments: _____

_____

24. Are premium variable-speed drive (VSD) motors used on the HVAC equipment? VSD motors save energy used on condenser and evaporator fans.

☐ ☐ ☐

Comments: _____

_____

25. Has radiant heating been considered for nonpublic areas such as warehouses and garages? Radiant heating warms objects instead of the air, and requires less fuel.

☐ ☐ ☐

Comments: _____

_____

26. Does the building have the proper levels of insulation? While the majority of insulation in buildings is for thermal purposes, the term also applies to acoustic insulation, fire insulation, and impact insulation (e.g., for vibrations caused by industrial applications).

☐ ☐ ☐

Comments: _____

_____

27. Have trade-offs been calculated between the extra cost of insulation and the current and estimated future cost of energy?

☐ ☐ ☐

Comments: _____

_____

C. Electrical Systems

1. Has the electrical system been planned for flexibility using floor duct trays and cable trays in ceilings in addition or in place of conventional conduits in the floor and walls?

☐ ☐ ☐

Comments: _____

_____

2. Are convenience outlets provided for standard electrical equipment: floor vacuums, scrubbers, polishers, clocks, computer terminals, and audiovisual equipment?

☐ ☐ ☐

Comments: _____

_____

	YES	NO	N/A

3. Are outlets away from walls and pillars and are they flush floor mounted and capped? Floor monuments are not recommended. ☐ ☐ ☐

   Comments: _____

   _____

4. Instead of fixed floor outlets, has a raised floor system been considered in all areas that may need electrical or communications relocation over the life of the project? ☐ ☐ ☐

   Comments: _____

   _____

5. Are additional empty conduits run to areas in the library that may require electrical power or communications equipment in the future? ☐ ☐ ☐

   Comments: _____

   _____

6. Is there at minimum a 1-inch dedicated conduit with a homerun from each and every data outlet to the telecommunications room? Another ¾-inch conduit is required for power. ☐ ☐ ☐

   Comments: _____

   _____

7. Do automated systems have dedicated data lines? ☐ ☐ ☐

   Comments: _____

   _____

8. Is the "sweep" on conduit runs gradual enough to accommodate fiber optic and coaxial cables? ☐ ☐ ☐

   Comments: _____

   _____

9. Does every public seat in the library have access to a duplex receptacle for power, and data communications and/or telephone outlets? All outlets should provide duplex power receptacles, and at least space for four data ports (coaxial, fiber, and twisted pair wires, with a box large enough to accommodate all four types of wire). ☐ ☐ ☐

   Comments: _____

   _____

10. Does every staff workstation have three to five duplex outlets, and data communications/telephone outlets? ☐ ☐ ☐

    Comments: _____

    _____

11. Are all cords and cables protected and out of sight? ☐ ☐ ☐

    Comments: _____

    _____

	YES	NO	N/A

12. Are dedicated lines provided for equipment requiring them? ☐ ☐ ☐

Comments: _____

_____

13. Before pouring floors, or enclosing conduit in walls, do the architect, contractor, and library staff "walk" the site to make sure that outlets are properly placed? ☐ ☐ ☐

Comments: _____

_____

14. Has a backup power source been considered for when the power is out? ☐ ☐ ☐

Comments: _____

_____

D. Lighting

1. Have the following lighting guidelines been followed?
   a) Maximize use of day lighting and integrate into electric lighting schemes. ☐ ☐ ☐
   b) Provide light-colored surfaces. Light-colored stack areas are critical. ☐ ☐ ☐
   c) Use task lighting at tables. ☐ ☐ ☐
   d) Use occupancy sensors for switching fixtures whenever possible. ☐ ☐ ☐
   e) Use dimming systems that are coupled to the amount of daylight within the space. ☐ ☐ ☐
   f) Increase the reflectance of walls (within contrast ratios). ☐ ☐ ☐
   g) Reduce glare by correctly choosing and placing fixtures. ☐ ☐ ☐
   h) Specify a minimum number of different lamp types at the building and site. ☐ ☐ ☐
   i) Reduce the number of decorative and display lights. ☐ ☐ ☐
   j) Provide a combination of lighting types. Include both general diffuse (indirect) and direct lighting. ☐ ☐ ☐
   k) Avoid large brightness ratios. Because brightness is a function of reflectance and illumination, the brightness level is controllable through good design. ☐ ☐ ☐
   l) Provide fixture locations that allow easy lamp replacement. Staff should not have to move furniture and equipment and bring in a scaffold in order to re-lamp light fixtures. ☐ ☐ ☐

   Comments: _____

   _____

2. Is the lighting energy- and cost-efficient? ☐ ☐ ☐

   Comments: _____

   _____

3. Will the building be designed to take advantage of natural light? Care must be taken to not cause glare or inconsistent light for users as well as not to damage books and furniture. ☐ ☐ ☐

   Comments: _____

   _____

| | YES | NO | N/A |

4. Have the advantages of natural light been considered?

   a) Enhanced visual quality. ☐ ☐ ☐

   b) Connection to nature. ☐ ☐ ☐

   c) Reduced energy cost. ☐ ☐ ☐

   d) Reduced HVAC. ☐ ☐ ☐

   e) Less use of energy resources. ☐ ☐ ☐

   Comments: _____

   _____

5. If day lighting is used, can it be controlled by window coverings, tinted glass, or other special glazing? ☐ ☐ ☐

   Comments: _____

   _____

6. When natural lighting is used, is it designed to eliminate glare and "hot spots" of intense light and/or heat? ☐ ☐ ☐

   Comments: _____

   _____

7. Can all interior and exterior lights be controlled from one location? ☐ ☐ ☐

   Comments: _____

   _____

8. Can staff operate a light control at staff entrances, allowing adequate illumination before arriving at the main control point for interior lighting? ☐ ☐ ☐

   Comments: _____

   _____

9. Are light switches located where they can be easily and logically accessed, not behind door swings or large pieces of equipment? ☐ ☐ ☐

   Comments: _____

   _____

10. Is the building's night lighting adequate to allow observation of the library's interior through outside windows? ☐ ☐ ☐

    Comments: _____

    _____

11. Are rheostat (dimmer) controls available at individual workstations to permit local adjustment to user needs? ☐ ☐ ☐

    Comments: _____

    _____

12. Are the following lighting levels maintained?

    a) Reading areas. 50 foot-candles average, measured horizontally at desktop, and augmented with task lighting carrels and table where appropriate. ☐ ☐ ☐

    b) Stacks. 20 foot-candles minimum sustained uniformly at floor level. ☐ ☐ ☐

c) Conference or study rooms. 30 to 40 foot-candles average measured
horizontally at desktop.

☐ ☐ ☐

d) Staff areas. 50 foot-candles average on desks or worktables measured
horizontally at desktop.

☐ ☐ ☐

e) Large meeting or community rooms. 40 foot-candles average with all lights on,
and with separately controlled lighting for the podium or front of the room.

☐ ☐ ☐

f) Parking lot. 0.6 foot-candles minimum measured horizontally on pavement,
to achieve a 4:1 average to minimum ratio, and with no spill light on adjacent
properties. Lighting must be sensitive to neighbors, have a higher illumination
level adjacent to the building and paths, and have a flexible control system that
can be adjusted by staff.

☐ ☐ ☐

Comments: _____

_____

13. Are ambient and task lights on timers or motion detectors in closed stacks, offices,
and/or public areas?

☐ ☐ ☐

Comments: _____

_____

14. Can ambient lighting be dimmed or brightened according to need?

☐ ☐ ☐

Comments: _____

_____

15. Is lighting zoned so various areas can be dimmed or brightened independently?

☐ ☐ ☐

Comments: _____

_____

16. Is flexible, timed programming available for each lighting zone?

☐ ☐ ☐

Comments: _____

_____

17. Do switch labels identify light zones?

☐ ☐ ☐

Comments: _____

_____

18. Can daylight be used as a source of lighting?

☐ ☐ ☐

Comments: _____

_____

19. Are computer monitors and other video screens shielded from direct sunlight or glare?

☐ ☐ ☐

Comments: _____

_____

20. Can lighting be easily moved if furniture, shelving, or equipment is moved?

☐ ☐ ☐

Comments: _____

_____

	YES	NO	N/A

21. Are components of the lighting system easily replaced and maintained? ☐ ☐ ☐

Comments: _____

_____

22. Are exterior lighting fixtures of vandal-resistant construction? ☐ ☐ ☐

Comments: _____

_____

23. Do exterior lighting fixtures have durable finishes to protect them from weather? ☐ ☐ ☐

Comments: _____

_____

24. Has the number of different lamp types been minimized to simplify maintenance
and lamp stocking? ☐ ☐ ☐

Comments: _____

_____

25. Are replacement lamps:
a) Easily accessible? ☐ ☐ ☐
b) Reasonably priced? ☐ ☐ ☐

Comments: _____

_____

26. Is there an emergency lighting system? ☐ ☐ ☐

Comments: _____

_____

E. Plumbing and Restrooms

1. Do all plumbing and restroom facilities meet the ADA guidelines described in chapter 7? ☐ ☐ ☐

Comments: _____

_____

2. Are restrooms constructed according to local building codes? ☐ ☐ ☐

Comments: _____

_____

3. Are restrooms and drinking fountains located near stairs, elevators, and other
permanent installations? The reason why this is desirable is because those areas
will not move and all immovable areas will be located together in a core. ☐ ☐ ☐

Comments: _____

_____

4. Are restrooms built above the level of the sewer system? ☐ ☐ ☐

Comments: _____

_____

	YES	NO	N/A

5. Does the number of sinks, toilets, and urinals meet local codes? ☐ ☐ ☐

   Comments: _____

   _____

6. Does the design of the restrooms accommodate one-third more toilets for women than men? ☐ ☐ ☐

   Comments: _____

   _____

7. Are the majority of toilets wall-hung to facilitate cleaning? ☐ ☐ ☐

   Comments: _____

   _____

8. Are some toilets floor mounted to accommodate plus-sized users? ☐ ☐ ☐

   Comments: _____

   _____

9. Are the toilets low-flow to conserve water? ☐ ☐ ☐

   Comments: _____

   _____

10. Have waterless urinals been considered? ☐ ☐ ☐

    Comments: _____

    _____

11. Have the best-quality fixtures and accessories been selected? The best-quality fixtures should be selected because there usually is a limited or no budget for replacing equipment, and because of the heavy use the fixtures will receive. ☐ ☐ ☐

    Comments: _____

    _____

12. Are the restrooms:

    a) Well ventilated (including fans)? ☐ ☐ ☐
    b) Well lighted? ☐ ☐ ☐
    c) Soundproof? ☐ ☐ ☐
    d) Vandal-resistant, especially the wall and stall surfaces? ☐ ☐ ☐

    Comments: _____

    _____

13. Are there provisions for:

    a) Toilet paper? ☐ ☐ ☐
    b) Soap? ☐ ☐ ☐
    c) Trash receptacles? ☐ ☐ ☐
    d) Towel dispensers or hand dryers? ☐ ☐ ☐
    e) Sanitary napkin dispensers? ☐ ☐ ☐
    f) Other? ☐ ☐ ☐

    Comments: _____

    _____

	YES	NO	N/A

14. Are dispensers planned and mounted to accommodate a change of vendors without damaging wall surfaces? ☐ ☐ ☐

Comments: _____

_____

15. Are there shelves for holding books and papers? ☐ ☐ ☐

Comments: _____

_____

16. Is there lockable closet/room storage for supplies? ☐ ☐ ☐

Comments: _____

_____

17. Are diaper-changing facilities available in all restrooms? ☐ ☐ ☐

Comments: _____

_____

18. Are restrooms available for transgender staff and patrons? Unrestricted restroom access and use according to an employee's full-time gender presentation should be available. ☐ ☐ ☐

Comments: _____

_____

19. Do restrooms meet the "Best Practices for Transgender Restroom Access," a new guide issued by the Occupational Safety and Health Administration in June 2015?

   a) Is there a written policy in place ensuring that all employees have access to appropriate facilities by allowing them to use restrooms that correspond with their gender identity? ☐ ☐ ☐

   b) Are there single-occupancy gender-neutral facilities? ☐ ☐ ☐

   c) Do multiple-occupant gender-neutral restroom facilities have a lockable single-occupant stall? ☐ ☐ ☐

   d) Are employees free to use the restroom of their choice without providing medical or legal documentation? ☐ ☐ ☐

   e) Are staff and customers free not to use a segregated facility apart from other employees due to their gender identity or transgender status? ☐ ☐ ☐

Comments: _____

_____

F. Elevators and Escalators

1. Are elevators/escalators located away from quiet areas? ☐ ☐ ☐

Comments: _____

_____

	YES	NO	N/A

2. Are there separate elevators for the public, staff, and/or freight? □ □ □

Comments: _____

_____

3. Do elevators/escalators meet ADA codes? □ □ □

Comments: _____

_____

4. Do elevators/escalators meet all local codes? □ □ □

Comments: _____

_____

5. Will the elevator/escalator system be designed so that routine maintenance will have minimal impact on library operations? □ □ □

Comments: _____

_____

G. Internet of Things to Monitor and Control Building Systems

1. What is the Internet of Things (IoT)? The IoT is the concept of basically connecting any device with an on and off switch to the Internet (and/or to each other). This includes everything from cell phones, coffeemakers, washing machines, headphones, lamps, wearable devices, and almost anything else you can think of. This also applies to components of machines, for example a jet engine of an airplane or the drill of an oil rig. □ □ □

Comments: _____

_____

2. Are building systems part of the IoT? Systems that control such building functions as heating, ventilation, and air conditioning (HVAC); security; refrigeration; and lighting have historically operated as stand-alone entities. □ □ □

Comments: _____

_____

3. What types of building systems may be controlled by the IoT?

   a) Will lighting be part of the library's IoT? For example, light fixtures may have an IP address and five sensors—all of them wired only to Ethernet cables. (They'll use "power over Ethernet" technology to deliver the juice to each fixture as well as data.) The fixtures include a light sensor to dim the LEDs during the day, and a motion detector that covers the area directly beneath each light and turns the light off when no one is there. □ □ □

   b) Will HVAC be part of the library's IoT? For example, HVAC systems in library buildings are often connected via building automation systems. The IoT entails a shift to more open communication and connection between all devices and systems within a building. □ □ □

c) Will Security—access control, intruder alarms, and video surveillance be part of the library's IoT? IP-enabled sensors lay at the heart of IoT as they sense and communicate the data needed to allow other devices to automatically make evaluations and take the necessary actions. Network cameras and video will be one of the most important sensors within the IoT because its data is so valuable to many other sensors and devices. ☐ ☐ ☐

d) Will fire and safety be part of the library's IoT? In addition to fire detection and alarm systems, additional capabilities such as incident management may be part of the system. ☐ ☐ ☐

e) Will energy be part of the library's IoT? The IoT helps building managers identify trends that can drive more efficient energy usage. For instance, software can monitor the real-time energy consumption of an HVAC, compare it against predicted usage, and provide the results visually. ☐ ☐ ☐

f) Will water management be part of the library's IoT? The key fixtures approach uses the premise that a building is water-efficient if the fixtures within the building are efficient. Water-efficient toilets, urinals, lavatories, showers, and food service equipment are targeted and designed (or replaced in existing buildings) for the highest level of water savings, and information about water use in the fixtures may be obtained by sensors. ☐ ☐ ☐

g) Will building access and security be part of the library's IoT? This includes connected building-access systems, video surveillance systems, and incident reporting systems. ☐ ☐ ☐

h) Will elevators and escalators be part of the library's IoT? Thousands of sensors and systems in its elevators and escalators that monitor everything from motor temperature to shaft alignment, cab speed, and door functioning are gathered and combined into a single dashboard that serves up two basic types of data: alarms that indicate an immediate issue, and events, which are stored and used for management. ☐ ☐ ☐

i) Will monitoring of equipment and maintenance needs be part of the library IoT? For example, smart fire extinguishers have been developed that monitor for tank pressure and space clearance and are able through Bluetooth technology to notify facility managers when these checks are found not to be in compliance or the extinguisher is removed from its mount. ☐ ☐ ☐

Comments: _____

_____

4. What are the advantages of using the IoT in controlling buildings?

a) Traditionally each building system has been proprietary and separated from mainstream systems and standards. The IoT allows all the systems to communicate.

b) It is cheaper to operate all building systems with one control point rather than many.

c) Building systems may be controlled from remote locations.

d) In the case of libraries that have multiple locations, it is possible to control systems in all buildings from one location.

e) Concerns over obsolescence also play a role in encouraging libraries to consider IoT to standardize system communication. Aging, proprietary controllers—computers that receive data on the building's environment and issue commands to devices—will eventually outlive vendor support.

f) Most experts believe that energy costs are reduced by using an IoT method to control building systems.

g) Staff training may be reduced because building system operators only need to learn one system.

Comments: _____

_____

5. What is the biggest concern about the IoT being used in buildings? Security is a
    big issue that is oftentimes brought up. With billions of devices being connected
    together, what can people to do to make sure that their information stays secure?
    Privacy and data sharing are also a concern. ☐   ☐   ☐

Comments: _____

_____

6. If the library plans an addition and/or remodel, will they try to integrate all building
    systems or phase in the most needed systems into the IoT? As the older technology
    fails, managers can replace proprietary controllers with new ones that employ open
    communications protocols. Both the legacy and the open-protocol building automation
    systems can exist simultaneously on the same network. ☐   ☐   ☐

Comments: _____

_____

7. Can the library use social media like Twitter to:

a) Learn of maintenance needs? ☐   ☐   ☐

b) Learn of environmental, physical, or security changes within the library? ☐   ☐   ☐

c) "Smart" fire extinguishers have been developed that monitor for tank pressure
    and space clearance and are able through Bluetooth technology to notify
    facility managers when these checks are found not to be in compliance or the
    extinguisher is removed from its mount. ☐   ☐   ☐

Comments: _____

_____

13

Safety and Security

A. General

1. Have all local codes regarding the safety of the occupants, building, and contents been met?

   □ □ □

   Comments: _____

   _____

2. Do the security measures provide a benefit of increased customer and staff safety without projecting a negative "police state" image?

   □ □ □

   Comments: _____

   _____

3. Do all alarm systems meet local codes when furnishings and decorations are in place?

   □ □ □

   Comments: _____

   _____

4. If the building is located in an earthquake zone, are all seismic protection measures in place?

   □ □ □

   Comments: _____

   _____

5. Are safety and security systems interconnected through the IoT?

   □ □ □

   Comments: _____

   _____

	YES	NO	N/A

6. Is there a plan in place for dealing with dangerous or disruptive behavior by customers or staff? □ □ □

    Comments: _____

    _____

7. Is there a plan in place for evacuation or for taking shelter in the building? □ □ □

    Comments: _____

    _____

B. External Security

1. Does the building require fencing to control access to the property? □ □ □

    Comments: _____

    _____

2. Is there sufficient, tamper-proof security lighting? □ □ □

    Comments: _____

    _____

3. Can a person climbing trees, fences, or the building structure, and so on gain access to roofs, upper windows, and ledges? □ □ □

    Comments: _____

    _____

4. If HVAC and electrical systems are in sheds or buildings outside the main building, are security screens provided to keep people from entering? □ □ □

    Comments: _____

    _____

5. Does the landscaping contribute to security by providing barriers to unauthorized entry? □ □ □

    Comments: _____

    _____

6. Does the landscaping detract from security by allowing locations for people to lurk? □ □ □

    Comments: _____

    _____

7. Are all vulnerable access points (doors, windows, air vents, etc.) protected against illegal entry with:

    a) High-security locks and hinges? □ □ □
    b) Security glazing? □ □ □
    c) Barriers (fences, grilles)? □ □ □
    d) Alarm systems? □ □ □
    e) Security camera systems (wired or wireless)? □ □ □
    f) Lighting systems? □ □ □

    Comments: _____

    _____

	YES	NO	N/A

8. Does the intrusion alarm:

   a) Transmit to the police or security company? ☐ ☐ ☐

   b) Immediately notify library personnel of an event? ☐ ☐ ☐

   c) Have automatic reset? ☐ ☐ ☐

   d) Have manual override? ☐ ☐ ☐

Comments: _____

_____

9. Are exterior book drops theft- and tamper-proof? ☐ ☐ ☐

Comments _____

_____

C. Internal Security

1. Is there a materials theft-detection system with alarm? ☐ ☐ ☐

Comments: _____

_____

2. Does the alarm transmit to a control or circulation desk? ☐ ☐ ☐

Comments: _____

_____

3. Are windows and emergency exits wired to prevent illegal use? ☐ ☐ ☐

Comments: _____

_____

4. Is there an emergency lighting system? ☐ ☐ ☐

Comments: _____

_____

5. Are all emergency exits clearly marked with lighted signs? ☐ ☐ ☐

Comments: _____

_____

6. Are exhibits, rare book collections, and other valuable materials provided with secure rooms and/or cases? ☐ ☐ ☐

Comments: _____

_____

7. Is valuable equipment attached to fixtures with security hardware? ☐ ☐ ☐

Comments: _____

_____

8. Can patrons gain undetected access to nonpublic areas? ☐ ☐ ☐

Comments: _____

_____

	YES	NO	N/A

9. Are there secluded areas that require convex mirrors or closed circuit TV? ☐ ☐ ☐

   Comments: _____

   _____

10. Are there areas where patrons can be undetected at closing? ☐ ☐ ☐

    Comments: _____

    _____

11. Is there an after-hours motion-detector system in place? ☐ ☐ ☐

    Comments: _____

    _____

12. If the building has a security staff, is their desk/office in a prominent location in order to act as a deterrent? ☐ ☐ ☐

    Comments: _____

    _____

D. Fire Safety

1. Is the building protected by a fire detection system, including smoke detectors? ☐ ☐ ☐

   Comments: _____

   _____

2. Are there carbon monoxide (CO) detectors? Carbon monoxide is a colorless, odorless gas that's known as the "silent killer." ☐ ☐ ☐

   Comments: _____

   _____

3. Are smoke and CO detectors adequately distributed? ☐ ☐ ☐

   Comments: _____

   _____

4. Does the alarm transmit to a fire station or central alarm station? ☐ ☐ ☐

   Comments: _____

   _____

5. Are fire hoses and extinguishers adequately distributed and highly portable? ☐ ☐ ☐

   Comments: _____

   _____

6. Is the building located close to a fire hydrant? ☐ ☐ ☐

   Comments: _____

   _____

7. Is there a sprinkler system? ☐ ☐ ☐

   Comments: _____

   _____

	YES	NO	N/A

8. Is shelving equipped with top panels to protect contents from water damage? ☐ ☐ ☐

   Comments: _____

   _____

9. Are there areas that require a special fire suppression system:

   a) Multiple-level, open stacks? ☐ ☐ ☐
   b) Rare-book collections? ☐ ☐ ☐
   c) Information labs and computer commons? ☐ ☐ ☐

   Comments: _____

   _____

E. Disaster Planning

The American Library Association has a website providing a variety of resources to deal with disasters. It contains almost everything that a library needs to plan for or to handle a disaster: www.ala.org/advocacy/govinfo/disasterpreparedness.

FEMA (Federal Emergency Management Agency) has a guide (www.ready.gov/make-a-plan) for how to plan for emergencies. This guide provides recommendations for the development of plans not only to respond to an emergency, but also outlines how public educational institutions can plan for preventing, protecting against, mitigating the impact of, and recovering from these emergencies. The guide translates lessons learned from the administration's work on national preparedness to benefit from recent advances in the emergency planning field. The guide introduces new approaches to planning that include walking through different emergency scenarios to create a course of action for each objective the team is trying to accomplish.

1. Has the library prepared plans to manage natural and man-made disasters? ☐ ☐ ☐

   a) Natural disasters and hazards include floods, hurricanes, thunderstorms, tornadoes, extreme heat or cold, earthquakes, landslides, and fires.
   b) Man-made threats include shootings, biological threats, chemical threats, nuclear blasts, and radiation.

   Comments: _____

   _____

2. Does the plan have the following components:

   a) Introduction—stating the lines of authority and the possible events covered by the plan. ☐ ☐ ☐
   b) Actions to be taken if advance warning is available. ☐ ☐ ☐
   c) First response procedures, including who should be contacted first in each type of emergency, what immediate steps should be taken, and how staff or teams will be notified. ☐ ☐ ☐
   d) Emergency procedures with sections devoted to each emergency event covered by the plan. This will include what is to be done during the event, and the appropriate salvage procedures to be followed once the first excitement is over. Floor plans should be included to illustrate what might be taking place in different parts of the library. ☐ ☐ ☐
   e) A rehabilitation plan for getting the library back to normal after the event. ☐ ☐ ☐

f) Appendixes, which may include evacuation/floor plans; listing of emergency services; listing of emergency response team members and responsibilities; telephone tree; location of keys; fire/intrusion alarm procedures; listing of collection priorities; arrangements for relocation of the collections; listing of in-house supplies; listing of outside suppliers and services; insurance information; listing of volunteers; prevention checklist; record-keeping forms for objects moved in salvage efforts; and detailed salvage procedures. □ □ □

Comments: _____

_____

3. Are the plans integrated into a larger campus or community plan? □ □ □

Comments: _____

_____

4. Does the library have a building escape plan? How will it be communicated to customers and staff? □ □ □

Comments: _____

_____

5. Does the library have a public address system to provide staff and customers with emergency information? □ □ □

Comments: _____

_____

6. Is there an instruction manual for staff on how to shut off gas, electricity, and other utilities? □ □ □

Comments: _____

_____

7. Has staff been given hands-on training on how to shut off utilities in case of emergency? □ □ □

Comments: _____

_____

8. Does the library have a first aid kit? □ □ □

Comments: _____

_____

9. Does the library have an area to take a person who needs aid for medical or other reasons? This usually requires a cot or at least a recliner chair. □ □ □

Comments: _____

_____

10. Will the library be used as a shelter during a campus or community disaster? □ □ □

Comments: _____

_____

	YES	NO	N/A

11. Does the library have a backup plan for critical documents and records? ☐ ☐ ☐

    Comments: _____

    _____

12. Does the library have a plan to mitigate damage to wet books and documents? ☐ ☐ ☐

    Comments: _____

    _____

13. Does the library have a plan to mitigate the damage from fire, including reducing odor and cleaning soot-damaged materials? ☐ ☐ ☐

    Comments: _____

    _____

14. How will the library handle infrastructure outages including power outages, drops, surges, utility outages, telecommunications failure, corrupt data, and so on? ☐ ☐ ☐

    Comments: _____

    _____

15. Is there a plan in place to deal with sabotage to library systems by customers or staff? ☐ ☐ ☐

    Comments: _____

    _____

14

Maintenance of Library Buildings and Property

A. Regular Routine Maintenance Considerations

1. Has a maintenance checklist been prepared for the custodial crew as well as for library staff who are responsible for checking that the work has been accomplished?

   □ □ □

   Comments: _____

   _____

2. Are maintenance supplies selected that will not emit air contaminants during use and storage?

   □ □ □

   Comments: _____

   _____

3. Are supplies stored in sealed, clearly labeled containers?

   □ □ □

   Comments: _____

   _____

4. Are chemicals, chemical-containing wastes, and containers disposed of according to manufacturers' instructions?

   □ □ □

   Comments: _____

   _____

5. Have dust control procedures been established:
   a) Are there barrier mats at the entrances?   □ □ □
   b) Are high-efficiency vacuum bags used?   □ □ □
   c) Are feather dusters wrapped with a dust cloth?   □ □ □
   d) Are air return grilles and air supply vents cleaned?   □ □ □

   Comments: _____

   _____

231

6. Moisture, leaks, and spills

   a) Is the building checked for moldy odors at least once a week?   ☐ ☐ ☐

   b) Are ceiling tiles, floors, and walls checked for leaks and discolorations at least
once a week?   ☐ ☐ ☐

   c) Are bathrooms and sinks checked for leaks on a daily basis?   ☐ ☐ ☐

   d) Are windows, windowsills, and window frames free of condensate?   ☐ ☐ ☐

   Comments: _____

_____

7. Are appliances checked for odors and/or leaks?   ☐ ☐ ☐

   Comments: _____

_____

8. Is a pest control system in place?   ☐ ☐ ☐

   a) Are the building and grounds checked for pest evidence, entry points, food,
water, and harborage sites?   ☐ ☐ ☐

   b) Are all potential pest habitats in buildings and grounds inspected and
monitored?   ☐ ☐ ☐

   c) Have alternative methods for treating pests been identified before pesticides
are used?   ☐ ☐ ☐

   d) Has a spot-treatment (or bait, crack, and crevice applications) been used
whenever pests are detected? This will reduce the amount of pesticide that
will need to be used.   ☐ ☐ ☐

   e) Are records kept of pesticide use, and do the records meet all of the local and
state requirements?   ☐ ☐ ☐

   Comments: _____

_____

B. Building Materials

1. Are exterior walls constructed of durable and easily maintained materials?   ☐ ☐ ☐

   Comments: _____

_____

2. Are interior walls constructed of durable and easily maintained materials?   ☐ ☐ ☐

   Comments: _____

_____

3. Are the materials used externally and internally energy-efficient?   ☐ ☐ ☐

   Comments: _____

_____

4. Are the materials used durable and of good quality?   ☐ ☐ ☐

   Comments: _____

_____

	YES	NO	N/A

5. Is the building constructed of fire-resistant materials? ☐ ☐ ☐

   Comments: _____

   _____

6. Can locally abundant building materials be used in the construction? ☐ ☐ ☐

   Comments: _____

   _____

7. Are windows built to help protect against direct sunlight and glare? ☐ ☐ ☐

   Comments: _____

   _____

8. Have natural colors and finishes been used and colors that would quickly become outdated been avoided? ☐ ☐ ☐

   Comments: _____

   _____

9. Do the colors and finishes complement the character of the surrounding community? ☐ ☐ ☐

   Comments: _____

   _____

C. Graffiti and Security

1. If graffiti occurs, is there a program in place to remove it as soon as possible? ☐ ☐ ☐

   Comments: _____

   _____

2. Does the landscaping create a barrier to help protect against vandalism? ☐ ☐ ☐

   Comments: _____

   _____

3. Is vegetation such as clinging vines used to cover walls to discourage graffiti? ☐ ☐ ☐

   Comments: _____

   _____

4. Are planter boxes used to protect walls? ☐ ☐ ☐

   Comments: _____

   _____

5. Will landscaping develop a dense mass against a wall so there is no room for graffiti? ☐ ☐ ☐

   Comments: _____

   _____

6. Is the building protected with a special coating or type of paint that allows for easy graffiti removal? ☐ ☐ ☐

   Comments: _____

   _____

	YES	NO	N/A

7. Does the building exterior have a color and texture to discourage graffiti? ☐ ☐ ☐

   Comments: _____

   _____

8. Does the concrete used have a color or pigmentation to discourage graffiti? ☐ ☐ ☐

   Comments: _____

   _____

9. Is there security lighting to discourage graffiti? ☐ ☐ ☐

   Comments: _____

   _____

10. Are fixtures high enough on walls to protect them from vandalism? ☐ ☐ ☐

    Comments: _____

    _____

11. Are fixtures sturdy enough to protect them from vandalism? ☐ ☐ ☐

    Comments: _____

    _____

12. Are signs high enough off the ground to protect them from vandalism? ☐ ☐ ☐

    Comments: _____

    _____

13. Is masonry or stone being used to protect areas that are particularly vulnerable to graffiti? ☐ ☐ ☐

    Comments: _____

    _____

14. Is the entrance secure from theft, vandalism, and graffiti? ☐ ☐ ☐

    Comments: _____

    _____

15. Is the building well lit, with light directed toward vulnerable areas and walkways? ☐ ☐ ☐

    Comments: _____

    _____

D. Building Systems Preventive Maintenance

Preventive maintenance is based on the needs of the individual library, the age of the building, and the budget available for maintenance.

1. Roofing
   a) Has a roofing file been created to review warranty information? ☐ ☐ ☐
   b) Is the roof inspected at least twice a year, and after any severe storm? ☐ ☐ ☐
   c) Is unnecessary traffic kept off of the roof to avoid membrane damage? ☐ ☐ ☐

	YES	NO	N/A

d) Are roof drains cleared of debris regularly (roof warranties usually don't cover this)? ☐ ☐ ☐

e) Is the perimeter walked periodically to examine sheet metal, copings, and previously repaired sections? ☐ ☐ ☐

f) Are roof-to-wall connections checked, and flashings examined (at curbs and penetrations, etc.) for wrinkles and tearing? ☐ ☐ ☐

g) Are single-ply roofs checked to determine if they need to be re-caulked at the top of face-mounted termination bars (if needed)? ☐ ☐ ☐

h) Are bituminous roofs checked for splits in the stripping plies? ☐ ☐ ☐

i) Is the roof checked for moisture every year to detect wet insulation or leaks? ☐ ☐ ☐

Comments: _____

---

2. Heating, ventilation, and air conditioning (HVAC)

a) Is the system inspected at least twice a year at seasonal start-ups (usually at the change from heating to cooling)? ☐ ☐ ☐

b) Does a licensed technician provide the service? ☐ ☐ ☐

c) Are cooling towers serviced annually? Various services need to be done based on the manufacturer's requirements and service should be recorded. ☐ ☐ ☐

d) Are pump bearings lubricated at least annually? ☐ ☐ ☐

e) Are pump couplings checked for leaks as soon as they are observed? ☐ ☐ ☐

f) Are HVAC units checked when unusual noises are heard? ☐ ☐ ☐

g) Are air-handling unit filters cleaned or replaced at least quarterly, or more often based on filter condition? ☐ ☐ ☐

Comments: _____

---

3. Plumbing

a) Is the plumbing system checked for leaks or unusual noises as needed? ☐ ☐ ☐

b) Are the domestic water booster and circulation pump system bearings lubricated at least annually? ☐ ☐ ☐

c) Are couplings checked for leaks on a regular basis? ☐ ☐ ☐

d) Are domestic water heaters and boilers fire-tested periodically? ☐ ☐ ☐

e) Are sump and sewage ejection pumps checked and replaced on an as-needed basis? ☐ ☐ ☐

f) Plumbing fixtures and traps:

   i) Are drains checked for proper drainage at least once a week? ☐ ☐ ☐

   ii) Are floor drains checked for proper drainage at least once a week? ☐ ☐ ☐

   iii) Are toilets checked for proper flushing once a week? ☐ ☐ ☐

Comments: _____

---

4. Lighting

a) Is the lighting system inspected at regular intervals? ☐ ☐ ☐

b) Are luminaires that have transformers, control gear, or accessories, such as spread lenses, glare baffles, or color filters, routinely checked? ☐ ☐ ☐

c) Are exterior lights regularly checked to make sure cables aren't torn; all screws and hardware are in place and working, and gaskets are checked and replaced when worn to provide a better watertight seal? ☐ ☐ ☐

d) Are burned-out lamps replaced on a timely basis? Consider group relamping (to create your relamping schedule, calculate lamp life and how often lamps are used). Group relamping usually saves labor costs. ☐ ☐ ☐

e) Are lamps checked to ensure that each lamp has the same color temperature?

f) Is adjustable lighting re-aimed if necessary? ☐ ☐ ☐

g) Are lamps dusted on a regular schedule and lens surfaces cleaned to enhance lighting performance? ☐ ☐ ☐

Comments: _____

5. Other systems

a) Do licensed or manufacturer-authorized professional personnel inspect fire and life safety systems on a regular basis? ☐ ☐ ☐

b) Does a licensed electrician inspect electrical systems every 3 to 5 years? ☐ ☐ ☐

c) Is the fire protection system checked on a regular basis? This includes checking pressure gauges, making sure that agent cylinders are in the operable range; that system piping and conduit are properly anchored; and that system nozzles aren't obstructed and allow adequate flow of the suppression agent into the protected space. ☐ ☐ ☐

d) Are parking surfaces cleaned at least twice a year, with coating replacement every 10 years? ☐ ☐ ☐

e) Is the building structure power washed as needed? This is a cheap, easy way to protect the structure and delay rehabilitation projects. ☐ ☐ ☐

Comments: _____

E. Building Cleaning

Cleaning schedules are established by the needs and budget of the individual library, but the following are some suggested schedules.

1. Building entry

a) Are entry mats vacuumed daily? ☐ ☐ ☐

b) Are floors swept/dusted 3x/week? ☐ ☐ ☐

c) Are floors damp mopped weekly? ☐ ☐ ☐

d) Is glass cleaned daily? ☐ ☐ ☐

e) Are steps and landings cleaned and cleared daily? ☐ ☐ ☐

f) Are porticos cleaned and cleared daily? ☐ ☐ ☐

g) If an outside ramp exists, is it cleaned monthly? ☐ ☐ ☐

h) Are entryways cleaned as needed? ☐ ☐ ☐

Comments: _____

	YES	NO	N/A

2. Hallways and lobbies

a) Is the area adjacent to the entry door swept or dust mopped daily? ☐ ☐ ☐
b) Are floors spot mopped daily? ☐ ☐ ☐
c) Are hard-surface floors scrubbed with a cleaning machine weekly? ☐ ☐ ☐
d) Are fingerprints and smudges on door handles and light switches cleaned weekly? ☐ ☐ ☐
e) Are drinking fountains cleaned daily? ☐ ☐ ☐
f) Are trash receptacles emptied daily? ☐ ☐ ☐
g) Are recycle bins emptied daily? ☐ ☐ ☐
h) Are desks, lamps and lampshades, chair rungs, low mouldings, sills, pictures, frames, and partition tops dusted weekly? ☐ ☐ ☐
i) Is restorative maintenance done as needed on a general schedule? For example, heavily traveled carpets are cleaned twice a year and carpeting is replaced every ten years. ☐ ☐ ☐

Comments: _____

3. Floors

a) Has a schedule been established for vacuuming and mopping floors? A minimum of 3x/week is suggested. ☐ ☐ ☐
b) Are spills on floors cleaned as soon as possible after they occur? ☐ ☐ ☐
c) Are hard-surfaced floors stripped and refinished every two years? ☐ ☐ ☐
d) Are carpets deep cleaned every eighteen months? This may need to be done more often in highly traveled areas. ☐ ☐ ☐
e) If carpet tiles are used, are they replaced as needed? ☐ ☐ ☐
f) If carpet tiles are used, does the library have replacement carpet tiles in storage? ☐ ☐ ☐

Comments: _____

4. Reading rooms

All of the cleaning schedules for hallways and lobbies apply to reading rooms.

a) Are seats and chairs inspected for wear and cleaned, repaired, and replaced as needed? ☐ ☐ ☐
b) Are trash and debris left on tables and chairs removed daily? ☐ ☐ ☐
c) Are books left on tables and chairs returned to stacks daily? ☐ ☐ ☐
d) Are tables and chairs dusted and cleaned if necessary on a weekly schedule? ☐ ☐ ☐
e) Are pencil sharpeners emptied daily? ☐ ☐ ☐

Comments: _____

5. Stacks

a) Are there problems in the stacks disclosed when cleaning? Droppings might signal the presence of rodents or insects and mold or dampness could indicate a humidity problem or leaky pipe. ☐ ☐ ☐
b) Are HEPA (high efficiency particulate air) vacuums used to clean shelving units? The HEPA vacuum prevents the recirculation of dust back into the air. It also has adjustable suction and a soft brush for cleaning especially fragile materials. ☐ ☐ ☐

c) Is a major stacks cleaning done every five years? Periodic dusting or vacuuming of stacks should be done as needed. ☐ ☐ ☐

Comments: _____

_____

6. Auditoriums

All of the cleaning schedules for hallways and lobbies apply to auditoriums.

a) If drapes or curtains are used, are they inspected and cleaned as needed? ☐ ☐ ☐

b) Are seats and chairs inspected for wear and cleaned, repaired, and replaced as needed? ☐ ☐ ☐

Comments: _____

_____

7. Conference/meeting rooms

All of the cleaning schedules for hallways and lobbies apply to conference/meeting rooms.

a) Are chalk and white boards cleaned on a weekly basis? ☐ ☐ ☐

b) If telecommunications equipment is used, is it cleaned and disinfected on a daily basis? ☐ ☐ ☐

Comments: _____

_____

8. Carrels/small study rooms

All of the cleaning schedules for conference/meeting rooms apply to carrels/small study rooms.

Comments: _____

_____

9. Offices/cubicles

All of the cleaning schedules for conference/meeting rooms apply to offices/cubicles.

Comments: _____

_____

10. Break rooms/kitchenettes

a) Is the floor cleaned daily? ☐ ☐ ☐

b) Are countertops cleaned daily? ☐ ☐ ☐

c) Are sinks cleaned and disinfected daily? ☐ ☐ ☐

d) Are the refrigerator's visible sides cleaned and shined daily? ☐ ☐ ☐

e) Are the refrigerator's coils cleaned and vacuumed twice a year? ☐ ☐ ☐

f) Is the dishwasher wiped inside and out, and its front cleaned and shined weekly? ☐ ☐ ☐

g) Is the microwave wiped inside and out, and its front cleaned and shined weekly? ☐ ☐ ☐

h) Are cabinets (outside) cleaned and polished four times a year? ☐ ☐ ☐

i) Are kitchen furniture (tables, chairs, desks) cleaned and polished daily? ☐ ☐ ☐

j) Are trash receptacles emptied daily and liners replaced? ☐ ☐ ☐

k) Are recycle bins emptied daily? ☐ ☐ ☐

Comments: _____

_____

	YES	NO	N/A

11. Windows/window coverings

    a) Are windows washed twice a year? ☐ ☐ ☐

    b) Are window coverings inspected twice a year (or as needed), and if damaged, repaired or replaced? ☐ ☐ ☐

    Comments: _____

_____

12. Restrooms

    a) Are floors and surfaces cleaned, mopped, and sanitized daily using germicidal cleaners? ☐ ☐ ☐

    b) Are vanities cleaned and sanitized daily? ☐ ☐ ☐

    c) Are sinks cleaned and disinfected daily? ☐ ☐ ☐

    d) Are mirrors and brightwork cleaned and polished daily? ☐ ☐ ☐

    e) Are all toilets, urinals, and washbasins cleaned and sanitized daily? ☐ ☐ ☐

    f) Are walls and partitions cleaned weekly? ☐ ☐ ☐

    g) Are all paper, soap, and sanitary napkin dispensers refilled daily? ☐ ☐ ☐

    h) Are all trash basket liners replaced daily? ☐ ☐ ☐

    Comments: _____

_____

13. Staff responsibility for cleaning

    Personal items (even if they are owned by the institution) should be cleaned by the individuals who own them. These items may include:

    a) Individual personal computers ☐ ☐ ☐

    b) Office desks and personal property (pictures, awards, etc.) ☐ ☐ ☐

    c) Clothing, cups, briefcases, and so on ☐ ☐ ☐

    Comments: _____

_____

F. Custodial Facilities

1. Is there adequate locking storage space allocated for janitorial supplies, tools, maintenance equipment, and so on, on each floor? ☐ ☐ ☐

    Comments: _____

_____

2. Is a sink or running water available in the custodial room and is the floor sloped with a floor drain? ☐ ☐ ☐

    Comments: _____

_____

3. Is the custodial room located as centrally as possible? ☐ ☐ ☐

    Comments: _____

_____

	YES	NO	N/A

4. Is there a custodial clothes closet or locker? Does the door have a louver or vent? ☐ ☐ ☐

Comments: _____

_____

5. Is there a mop, broom, and brush rack? ☐ ☐ ☐

Comments: _____

_____

6. Is there a desk or worktable and tool storage area for minor repairs? ☐ ☐ ☐

Comments: _____

_____

7. Is the door wide enough for ease of moving equipment in and out of the space? ☐ ☐ ☐

Comments: _____

_____

8. Is the wall area around the sink made of a durable material to prevent water damage? ☐ ☐ ☐

Comments: _____

_____

G. Groundskeeper Facilities

1. Is there provision for secure storage of lawnmowers, snow blowers, and other equipment? ☐ ☐ ☐

Comments: _____

_____

2. Is there provision for adequate outside faucets and electrical outlets? ☐ ☐ ☐

Comments: _____

_____

3. Are faucets, irrigation equipment, and electrical outlets vandal-proof? ☐ ☐ ☐

Comments: _____

_____

H. Trash Enclosures

1. Is there adequate exterior space allocated for the storing of trash? ☐ ☐ ☐

Comments: _____

_____

2. Is the trash area easily accessible from the building and from the street for pickup? ☐ ☐ ☐

Comments: _____

_____

	YES	NO	N/A

3. Is there adequate space allowed for garbage truck maneuvering and/or turnaround? ☐ ☐ ☐

    Comments: _____

    _____

4. Is the garbage bin hidden or camouflaged from public view with shrubs or a decorative wall? ☐ ☐ ☐

    Comments: _____

    _____

5. Is the area secure from scavenging?

    Comments: _____

    _____

I. Betterments and Improvements

1. Does the library have a policy for betterments and improvements? Betterments and improvements are minor capital improvement projects with a spending threshold set by the university/community, school district, and so on. Betterments and improvements extend the life, increase the productivity, or significantly improve the safety (for example, asbestos removal) of the building as opposed to repairs and maintenance that either restore the asset to, or maintain it at, its normal or expected service life. ☐ ☐ ☐

    Comments: _____

    _____

2. In order to qualify as a betterment and improvement, do the benefits from the project need to extend for a stated number of years? ☐ ☐ ☐

    Comments: _____

    _____

3. Does the library have a procedure for requesting betterments and improvements from the library's parent institution (campus, city, etc.)? ☐ ☐ ☐

    Comments _____

    _____

4. Does the parent institution or board have a policy in place for approving requests? ☐ ☐ ☐

    Comments: _____

    _____

5. Are projects to correct safety issues given first priority? ☐ ☐ ☐

    Comments _____

    _____

6. Are there minor capital improvement projects that are ineligible? For example, some institutions prohibit projects dealing with maintenance or repairs. ☐ ☐ ☐

    Comments: _____

    _____

15

Building Occupancy and Post-Occupancy Evaluation

A. Building Acceptance

1. Has a library representative, the contractor, the architect, and other interested parties conducted a building inspection to develop a punch list? A punch list is developed during a walk through the building at the end of construction to determine what is not finished. A contractor may develop a preliminary punch list and the walk-through refines and usually increases the list of items to be completed. ☐ ☐ ☐

    Comments: _____

    _____

2. After the punch list walk-through is completed, is there:
    a) A list of what needs to be fixed? ☐ ☐ ☐
    b) A list of what is missing, or what needs to be ordered? ☐ ☐ ☐
    c) One person who documents problems or suggestions? ☐ ☐ ☐
    d) Is the information communicated to all the staff? ☐ ☐ ☐
    e) A plan and procedure to clear items on the punch list? ☐ ☐ ☐

    Comments: _____

    _____

3. When the building is accepted by the owner (library), will the architect and contractor provide:
    a) "As built" drawings that document what was really built? Often during construction, what is built may vary from the original drawings and corrections made in the as built drawings. It is vital that the library have these drawings for future work in the building. ☐ ☐ ☐

b) Shop drawings from the contractor to the architect documenting what was built? These drawings provide detail that is not in the original set of drawings from the architect. Shop drawings are drawings, diagrams, schedules, and other data specially prepared for the building project by the contractor or a subcontractor, manufacturer, supplier, or distributor to illustrate some portion of the work. The shop drawing normally shows more detail than the construction documents. It is drawn to explain the fabrication and/or installation of the items to the manufacturer's production crew or contractor's installation crews. The contractor is obligated by the contract documents to submit shop drawings, product data, and samples for certain parts of the work to the architect.   ☐ ☐ ☐

Comments: _____

_____

4. Has a keying and access system been decided?   ☐ ☐ ☐

Comments: _____

_____

5. Will the contractor provide a set of keys to the library after building acceptance? The library should have consulted with the architect and contractor early in construction to determine what keys will be required.   ☐ ☐ ☐

Comments: _____

_____

6. Has a room numbering system been decided?   ☐ ☐ ☐

Comments: _____

_____

B. Certificate of Occupancy

When construction work is complete, the library (or its parent agency) requires a certificate of occupancy (CO). This is another step that may delay the opening of the library building, and it is prudent to have building officials involved in all phases of the project from planning to construction.

1. Is all the work described when the building permit was issued been completed and has the building had a final inspection by the local governing agency?   ☐ ☐ ☐

Comments: _____

_____

2. Has the building received a certificate of completion or occupancy by the local agency's building official? A CO is a document issued by a building official certifying that all or a designated portion of a building complies with the provisions of the building code, and permits occupancy for the buildings designated use. A library cannot occupy the space without a certificate of occupancy.   ☐ ☐ ☐

Comments: _____

_____

3. Has the library received a temporary certificate of occupancy? A temporary certificate of occupancy may be issued when the building official finds that no substantial hazard will result from occupancy of any structure or portion of a structure before all work is completed. A temporary certificate may allow movement of equipment and books into the building. A CO is required upon completion of the remainder of the work in order to allow occupancy of the building. ☐ ☐ ☐

   Comments: _____

   _____

4. Have the following items been inspected and signed off by the proper building officials?
   a) Engineering? ☐ ☐ ☐
   b) Electrical? ☐ ☐ ☐
   c) Plumbing? ☐ ☐ ☐
   d) Fire? ☐ ☐ ☐
   e) Water/sewer? ☐ ☐ ☐
   f) Backflow/cross-connection control? ☐ ☐ ☐
   g) Landscaping? ☐ ☐ ☐
   h) Mitigation monitoring coordination? ☐ ☐ ☐

   Comments: _____

   _____

5. Once the above inspection items are approved, the following items are also required in order to obtain a CO:
   a) Has a copy of the signed inspection record card been received and filed? ☐ ☐ ☐
   b) Has an application been submitted to the inspections office for the issuance of the CO? ☐ ☐ ☐
   c) Pending final approval from the building inspector, the building official will release the appropriate utilities. ☐ ☐ ☐

   Comments: _____

   _____

C. Getting Ready for Occupancy

1. Will there be any organizational changes in the new building, and if so, have they been explained to staff? ☐ ☐ ☐

   Comments: _____

   _____

2. Has the library's budget been adjusted to accommodate the new building (additional staff, utilities, etc.)? ☐ ☐ ☐

   Comments: _____

   _____

3. Will the library be required to change rules and regulations as a result of the new building? ☐ ☐ ☐

   Comments: _____

   _____

4. Have VIP and staff tours been scheduled during the planning and construction process to get people involved and energized? ☐ ☐ ☐

   Comments: _____

   _____

5. Have virtual tours been scheduled for the staff so that they can envision where they will be working? ☐ ☐ ☐

   Comments: _____

   _____

6. Has staff been "psychologically prepared" for the new working conditions?

   a) Have they been involved in the initial planning? ☐ ☐ ☐
   b) Have they reviewed plans and models in order to make suggestions? ☐ ☐ ☐
   c) Have they been given an opportunity to "personalize" their workstations and work areas? ☐ ☐ ☐
   d) Have they had the opportunity to tour the building while it is under construction? ☐ ☐ ☐

   Comments: _____

   _____

7. Has the anticipated increased use of the facility been planned for? ☐ ☐ ☐

   Comments: _____

   _____

8. Has all the furniture and equipment been ordered so that it will arrive prior to the library's scheduled opening? If not, opening of the library may need to be postponed. ☐ ☐ ☐

   Comments: _____

   _____

D. Moving

1. Will the library employ a library-moving specialist, or will the library move with internal resources? ☐ ☐ ☐

   Comments: _____

   _____

2. Can the move to the new space be scheduled during the time when demand for library services is at its lowest level of activity? ☐ ☐ ☐

   Comments: _____

   _____

3. Has the library allowed adequate time from the ending of services in the old building to the establishment of services in the new building? Libraries often underestimate the time required, especially in larger building projects. ☐ ☐ ☐

   Comments: _____

   _____

	YES	NO	N/A

4. Will the old library need to be closed in order to move to the new space? ☐ ☐ ☐

   Comments: _____

   _____

5. If the building project is an expansion, will the library be able to operate during construction of the addition? This sometimes is not determined until the contractor starts construction. ☐ ☐ ☐

   Comments: _____

   _____

6. If the library needs to close, what is the window of time it can remain open before it needs to be closed to move into the new space? ☐ ☐ ☐

   Comments: _____

   _____

7. What approval from the university, college, school district, library board, or city council is required to approve the date of closing and the duration of time the library will be closed? ☐ ☐ ☐

   Comments: _____

   _____

8. How will library customers be notified about the closing, and what new procedures will be required for interim service? ☐ ☐ ☐

   Comments: _____

   _____

9. How much of the existing collection, stacks, furniture, and equipment will be moved to the new building? ☐ ☐ ☐

   Comments: _____

   _____

10. Has the amount to be moved been calculated and measured so that it will fit into the new space? This includes books and media, furniture, and anything that will be part of the new building. ☐ ☐ ☐

    Comments: _____

    _____

11. Will the items being moved be cleaned before the move? ☐ ☐ ☐

    Comments: _____

    _____

12. Have timetables and schedules been made to plan all stages of the move? ☐ ☐ ☐

    Comments: _____

    _____

13. Have plans been made as to what to do with the furniture, fixtures, and equipment that are not moved? ☐ ☐ ☐

Comments: _____

_____

14. Have plans been made as to what to do with the old library? ☐ ☐ ☐

Comments: _____

_____

E. Post-Occupancy Evaluation

1. Will a post-occupancy evaluation be conducted? This is the process of diagnosing the technical, functional, and behavioral aspects of a completed building in order to accumulate information for future programming and design activities. ☐ ☐ ☐

Comments: _____

_____

2. When will the post-occupancy evaluation be conducted? Usually it takes place one year from the date when the certificate of occupancy is issued. One year allows four seasons of use. ☐ ☐ ☐

Comments: _____

_____

3. Was the building completed on time? ☐ ☐ ☐

Comments: _____

_____

4. Was the building completed within budget? ☐ ☐ ☐

Comments: _____

_____

5. Did the building meet the library's building program? If not, what caused variance from the building program? ☐ ☐ ☐

Comments: _____

_____

6. Does the staff like the building, and if not, what can be changed to solve the problem? ☐ ☐ ☐

Comments: _____

_____

7. Do library customers like the building? Sometimes a survey is conducted to see customer concerns and if possible to make changes to the new building. ☐ ☐ ☐

Comments: _____

_____

YES | NO | N/A

8. Has the library had problems in maintaining the building? If so, what can be done to mitigate maintenance problems? □ □ □

Comments: _____

_____

9. Did the architect provide all the services specified in his contract? □ □ □

Comments: _____

_____

10. Was the architect responsive to the needs of the client? □ □ □

Comments: _____

_____

11. Did the architect adequately represent the client in negotiations with all of the stakeholders? □ □ □

Comments: _____

_____

12. Did the contractor maintain a clean and safe job site? □ □ □

Comments: _____

_____

13. Did the contractor identify problems in the drawings and/or specifications during the project? □ □ □

Comments: _____

_____

14. During the shakedown period (usually the one-year warranty period after the building is accepted by the owner), were errors and/or omissions in the new building brought to the attention of the architect and contractor? □ □ □

Comments: _____

_____

15. Were all errors and/or omissions resolved to the owner's satisfaction during the shakedown period? □ □ □

Comments: _____

_____

16. If errors and omissions occurred, was it possible to determine who was responsible?

   a) Owner? □ □ □
   b) Architect? □ □ □
   c) Contractor? □ □ □
   d) Undetermined or out of control of any of the players? □ □ □

Comments: _____

_____

17. In the final evaluation, was the building project:

   a) Planned and designed to reinforce the library as a center of the campus or community? ☐ ☐ ☐

   b) Designed to provide for comfort and health, safety and security to the campus or community? ☐ ☐ ☐

   c) Designed to make effective use of all available resources? ☐ ☐ ☐

   d) Designed to address changing library needs over time by permitting flexibility and adaptability? ☐ ☐ ☐

Comments: _____

_____

16

Groundbreaking and Dedication Ceremonies

A. Planning

1. Does the parent institution (university, city, etc.) have procedures and protocols in place that must be followed for groundbreakings and grand openings/dedication ceremonies?

   ☐ ☐ ☐

   Comments: _____

   _____

2. Are there procedures in place for a groundbreaking ceremony? A groundbreaking is a ceremony designed to symbolically "break ground" at a location that has officially been approved for new construction and to honor those that have made it possible.

   ☐ ☐ ☐

   Comments: _____

   _____

3. Are there procedures in place for a grand opening/dedication ceremony? A dedication/grand opening is a ceremony that commemorates the completion and opening of a new building or major renovation and may symbolically "dedicate" the building in honor of a person or persons that have made the construction or renovation possible.

   ☐ ☐ ☐

   Comments: _____

   _____

4. Has a committee been formed by the parent institution to plan the event?

   ☐ ☐ ☐

   Comments: _____

   _____

5. Has the committee considered and made recommendations concerning the following planning items?

    a) Goals of the event?   ☐ ☐ ☐

    b) Program content?   ☐ ☐ ☐

    c) Printed program?   ☐ ☐ ☐

    d) Who will emcee the event?   ☐ ☐ ☐

    e) Acknowledgments to be made during the event and in the program?   ☐ ☐ ☐

    f) Site for event and backup for inclement weather?   ☐ ☐ ☐

    g) Speakers and presentation order?   ☐ ☐ ☐

    h) Seating, both reserved and open?   ☐ ☐ ☐

    i) Ushers?   ☐ ☐ ☐

    j) Gifts or/token for key participants?   ☐ ☐ ☐

    k) Decorations and setup on stage?   ☐ ☐ ☐

    l) Publicity?   ☐ ☐ ☐

    m) Flags?   ☐ ☐ ☐

    n) Traffic control?   ☐ ☐ ☐

    o) VIP parking?   ☐ ☐ ☐

    p) Extra custodial care?   ☐ ☐ ☐

    q) Ribbon cutting, banner, or ceremony?   ☐ ☐ ☐

    r) Signs?   ☐ ☐ ☐

    s) Tours?   ☐ ☐ ☐

    t) Refreshments?   ☐ ☐ ☐

    u) Music?   ☐ ☐ ☐

    Comments: _____

    _____

6. Is timing of the event tied to the schedule of one or a few key people? For example, calendars of the provost, mayor, key donor, and so on may determine the date and time.   ☐ ☐ ☐

    Comments: _____

    _____

7. Have the key participants been informed of and agreed to the date and time of the event?   ☐ ☐ ☐

    Comments: _____

    _____

8. Is the date and time convenient to all of the people who may be interested in the event?   ☐ ☐ ☐

    Comments: _____

    _____

9. Are there other events occurring on campus or in the community that may conflict with the event? For example, don't schedule the opening of the new life sciences library at the University of Michigan on the date of the Michigan/Ohio State football game.   ☐ ☐ ☐

    Comments: _____

    _____

	YES	NO	N/A

10. Have invitations to the event been sent out in a timely manner? (Allow at least one month before the event.) ☐ ☐ ☐

   Comments: _____

   _____

11. Is there a procedure for determining attendance? ☐ ☐ ☐

   Comments: _____

   _____

12. Has one person been designated to act as the coordinator for the event? ☐ ☐ ☐

   Comments: _____

   _____

13. Has it been determined who will have speaking parts at the event? ☐ ☐ ☐

   Comments: _____

   _____

14. Do speakers know their time limits, and is there a way to keep them to the limits? ☐ ☐ ☐

   Comments: _____

   _____

15. Are devoted, talented people assigned to handle the various jobs required to make the event successful? ☐ ☐ ☐

   Comments: _____

   _____

16. Do all of the people working on the event know their roles and responsibilities? ☐ ☐ ☐

   Comments: _____

   _____

17. Has publicity been prepared and scheduled? ☐ ☐ ☐

   Comments: _____

   _____

18. Have press releases and informational packets been sent to the local media? ☐ ☐ ☐

   Comments: _____

   _____

19. Have the media been contacted and urged to cover the event? ☐ ☐ ☐

   Comments: _____

   _____

20. Has a media contact person been identified and listed in all publicity, including the contact's address and telephone number? ☐ ☐ ☐

   Comments: _____

   _____

	YES	NO	N/A

21. Is there someone responsible for making an audio and video history of the event? ☐ ☐ ☐

Comments: _____

_____

22. Is there someone responsible for gathering and keeping information from the event such as a guest book, gifts, and so on? ☐ ☐ ☐

Comments: _____

_____

23. Will the event be short, interesting, and focused? ☐ ☐ ☐

Comments: _____

_____

B. Event Checklist

1. Have street closures, parking, and traffic control been coordinated with the local law authorities? ☐ ☐ ☐

Comments: _____

_____

2. Will the site be inspected and cleaned up before the event? ☐ ☐ ☐

Comments: _____

_____

3. Will there be adequate signage indicating where attendees are to go? ☐ ☐ ☐

Comments: _____

_____

4. For groundbreakings, will there be "ceremonial shovels" and hard hats available? ☐ ☐ ☐

Comments: _____

_____

5. For dedications, will there be "ceremonial scissors" available? ☐ ☐ ☐

Comments: _____

_____

6. Has a source been found to provide:

   a) Tables? ☐ ☐ ☐
   b) Chairs? ☐ ☐ ☐
   c) Podium? ☐ ☐ ☐
   d) Barricades? ☐ ☐ ☐
   e) Public-address system? ☐ ☐ ☐
   f) Stage? ☐ ☐ ☐
   g) Flags? ☐ ☐ ☐
   h) Institutional seals? ☐ ☐ ☐

	YES	NO	N/A

i) Refreshments? ☐ ☐ ☐

j) Tablecloths, napkins, plates, silverware, and cups? ☐ ☐ ☐

k) Trash cans/bags? ☐ ☐ ☐

l) Plants or decorations? ☐ ☐ ☐

m) Bathrooms and toilet supplies? ☐ ☐ ☐

Comments: _____

_____

7. Have all of the dignitaries been invited?

a) University or college administration? ☐ ☐ ☐

b) Faculty senate president? ☐ ☐ ☐

c) Mayor? ☐ ☐ ☐

d) City council? ☐ ☐ ☐

e) Architect? ☐ ☐ ☐

f) Contractor? ☐ ☐ ☐

g) Project manager? ☐ ☐ ☐

h) Friends of the Library? ☐ ☐ ☐

i) Community groups? ☐ ☐ ☐

j) Library VIPs? ☐ ☐ ☐

Comments: _____

_____

8. Will name tags be available? ☐ ☐ ☐

Comments: _____

_____

9. Will souvenirs or commemorative items be given to attendees? ☐ ☐ ☐

Comments: _____

_____

10. Will VIPs receive special items? ☐ ☐ ☐

Comments: _____

_____

11. Will a guest book be available allowing event attendees to sign in? ☐ ☐ ☐

Comments: _____

_____

12. Have invitations been:

a) Designed? ☐ ☐ ☐

b) Printed? ☐ ☐ ☐

c) Checked and checked again for accuracy? ☐ ☐ ☐

d) Mailed at least one month before the event? ☐ ☐ ☐

e) Copies saved for the library's archives? ☐ ☐ ☐

Comments: _____

_____

	YES	NO	N/A

13. Has the program been:

    a) Designed? ☐ ☐ ☐

    b) Printed? ☐ ☐ ☐

    c) Checked and checked again for accuracy? ☐ ☐ ☐

    d) Copies saved for the library's archives? ☐ ☐ ☐

    Comments: _____

_____

14. Is there a master of ceremonies for the event, and how is that person selected? ☐ ☐ ☐

    Comments: _____

_____

15. Will speakers:

    a) Know and adhere to their time limit? ☐ ☐ ☐

    b) Provide the master of ceremonies with biographical information for introductions? ☐ ☐ ☐

    c) Provide copies of their remarks for the library's archives? ☐ ☐ ☐

    d) Know when to arrive and where to sit? ☐ ☐ ☐

    e) Know the proper attire to wear? ☐ ☐ ☐

    Comments: _____

_____

16. Will individuals be recognized for their contributions to the building project? This might include the architect, contractor, and so on. ☐ ☐ ☐

    Comments: _____

_____

17. Has music been arranged for the event? ☐ ☐ ☐

    Comments: _____

_____

18. Will there be a color guard for a national anthem/colors ceremony? ☐ ☐ ☐

    Comments: _____

_____

19. If bad weather has the potential to affect the event, is there an alternative plan available? ☐ ☐ ☐

    Comments: _____

_____

20. Will the site be cleaned after the event? ☐ ☐ ☐

    Comments: _____

_____

21. Will thank-you letters be sent to:

    a) Dignitaries? ☐ ☐ ☐

    b) Speakers? ☐ ☐ ☐

    c) Donors? ☐ ☐ ☐

    d) Volunteers? ☐ ☐ ☐

	YES	NO	N/A
e) Friends?	☐	☐	☐
f) Staff?	☐	☐	☐
g) Architects and construction crew?	☐	☐	☐
h) Others?	☐	☐	☐

Comments: _____
_____

22. Will someone be responsible for writing a report on what was learned from the
event? This will help in planning future events.   ☐   ☐   ☐

Comments: _____
_____

Bibliography

Aaronson, Lauren. "An iPod for Your Books." *Popular Science* 269, no. 2 (August 2007): 17.

Allen, Walter C. "Selected References." *Library Trends* 36, no. 2 (1987): 475–91.

American Association of School Librarians. "Learning Standards & Program Guidelines Implementation Toolkit." 2015. www.ala.org/aasl/advocacy/tools/toolkits/standards -guidelines.

American Institute of Architects. "ADA Standards 2010; Architectural Barriers Act and IBC + ANSI A117.1, 2012."

American Library Association. "Building Libraries and Library Additions: A Selected Annotated Bibliography: ALA Library Fact Sheet 11." *Library Fact Sheets*. 2015. www.ala.org/tools/libfactsheets/alalibraryfactsheet11.

Americans with Disabilities Act (ADA) U.S. Access Board. *Americans with Disabilities Act (ADA): Accessibility Guidelines for Buildings and Facilities*.

Anixter. "The Six Subsystems of a Structured Cabling System." 2015. www.anixter .com/en_us/resources/literature/techbriefs/the-six-subsystems-of-a-structured -cabling-system.html.

Association of College and Research Libraries (ACRL)/Library Leadership and Management Association (LLAMA). "Academic Library Building Design: Resources for Planning." *ACRL Wiki*. 2015. www.wikis.ala.org/acrl/index.php/Academic_Library _Building_Design:_Resources_for_Planning.

Bahr, Alice Harrison. "Library Buildings in a Digital Age: Why Bother?" *College & Research Libraries News* 61, no. 7 (2000): 590–91.

Ball, Steven J. "Design Thinking." *American Libraries* 39 (2008): 44–49.

Barista, Dave. "Designing for the Obese." *Building Design + Construction*, November 2005.

Bartle, Lisa. R. "Designing an Active Academic Reference Service Point." *Reference & User Services Quarterly* 38, no. 4 (1999): 395.

Baumann, Charles H. *The Influence of Angus Snead MacDonald and the Snead Bookstack on Library Architecture*. Metuchen, NJ: Scarecrow, 1972.

Bazillion, Richard J., and Connie Braun. *Academic Libraries as High Tech Gateways*. Chicago: American Library Association, 1995.

Becket, Margaret, and Henry B. Smith. "Designing a Reference Station for the Information Age." *Library Journal* (April 1986): 42–46.

Bennett, Scott, Geoffrey T. Freeman, American Institute of Architects, Bernard Frischer, Kathleen Burr Oliver, and Christina A. Peterson. *Library as Place: Rethinking Roles, Rethinking Space*. Washington, DC: Council on Library and Information Resources, 2005.

Bernheim, Anthony, and American Institute of Architects. *San Francisco Main Library: A Healthy Building*. IFLA Council and Conference. Barcelona: IFLA, 1993.

Bio-Link. "Bio-Link National Center." 2014. www.bio-link.org/home.

Bisbrouck, Marie-Françoise, Jérémie Desjardins, Céline Ménil, Florence Poncé, and François Rouyer-Gayette, eds. *Libraries as Places: Buildings for the 21st Century*. International Federation of Library Associations and Institutions. Paris: K. G. Saur (Munich), 2003.

Black, J. B., Janet Black, Ruth O'Donnell, and Jane Scheuerle. *Surveying Public Libraries for the ADA*. Tallahassee: Bureau of Library Development, Division of Library and Information Services, State Library of Florida, 1993.

Block, David. "Remote Storage in Research Libraries: A Microhistory." *Library Resources & Technical Services* 44, no. 4 (2000): 184–89.

Boaz, Martha. *A Living Library: Planning Public Library Buildings for Cities of 100,000 or Less*. Los Angeles: University of Southern California Press, 1957.

Boss, Richard W. "Automated Storage/Retrieval and Return/Sorting Systems." *Tech Notes*. www.pla.org/publications/technotes/technotes.html.

———. *Information Technologies and Space Planning for Libraries and Information Centers*. Boston: G. K. Hall, 1987.

———. *The Library Manager's Guide to Automation*. White Plains, NY: Knowledge Industry Publications, 1984.

———. *Telecommunications for Library Management*. White Plains, NY: Knowledge Industry Publications, 1985.

Bostick, Sharon L., and Bryan Irwin. *Designing an Academic Library as a Place and a Space: How a Robotic Storage System Will Create a Twenty-First Century Library Design*. Milan: International Federation of Library Systems, 2009.

Brawner, Lee, and Donald K. Beck, Jr. *Determining Your Public Library's Future Size*. Chicago: American Library Association, 1996.

Brec, R. L. "Playing 20 Questions." *San Diego Daily Transcript*, November 16, 1999, 9A.

Breeding, Marshall. *Library LANs: Case Studies in Practice and Application*. Westport, CT: Meckler, 1992.

Brown, Carol R. *Planning Library Interiors: The Selection of Furnishings for the 21st Century*. Phoenix, AZ: Oryx, 1995.

———. *Selecting Library Furniture: A Guide for Librarians, Designers, and Architects*. Phoenix, AZ: Oryx, 1989.

Brown-Sica, Margaret S. "Library Spaces for Urban, Diverse Commuter Students: A Participatory Action Research Project." *College & Research Libraries News* 73, no. 3 (2012): 14.

California Library Association. *Earthquake Preparedness Manual for California Libraries*. Sacramento: California Library Association, 1990.

Campbell, Anne L. "Magical Models." *Library Journal* 128, no. 3 (2003): 38–40.

Carlson, Scott. "An Anthropologist in the Library." *The Chronicle of Higher Education; Information Technology Section* 53, no. 50 (2007): 26.

———. "As Students Work Online, Reading Rooms Empty Out—Leading Some Campuses to Add Starbucks." *Chronicle of Higher Education* 48, no. 12 (2001): A35–A38.

Carroll, R. E. "Building a Library: The Librarian/Architect Relationship." *New Zealand Libraries* 45 (1987): 85–89.

Carvajal, Doreen. "Racing to Convert Books to Bytes." *New York Times*, December 9, 1999, 7.

Colorado WaterWise Council. "Xeriscape™ Colorado." 2008. www.xeriscape.org/planning article.html.

Cheek, Lawrence. "On Architecture: How the New Central Library Really Stacks Up." *Seattle Post-Intelligencer*, March 27, 2007, 1–3.

Ching, Francis D. K. *A Visual Dictionary of Architecture*. New York: Van Nostrand Reinhold, 1995.

City and County of San Francisco. *Industrial Areas Design Guideline*. San Francisco Planning Department. City and County of San Francisco (2001), 40.

Clancy, Heather. "How the Internet of Things Will Transform Building Management." *GreenBiz*. 2013. www.greenbiz.com/news/2013/03/27/internet-things-building -management.

Clark, Don. "Command Centers for Smart Homes." *Wall Street Journal*, January 5, 2015, 2.

Cohen, Elaine. "Library Facilities." *Bookmark* (Spring 1990): 210–12.

Cohen, Elaine, and Aaron Cohen. *Designing and Space Planning for Libraries: A Behavioral Guide*. New York: Bowker, 1979.

College and Research Libraries Task Force. "Standards for Libraries in Higher Education: A Draft." *C&RL News* 64, no. 5 (2003): 329–36.

Corban, Gaylan. "Ideal Archives Environment." E-mail, January 13, 1997.

Crooks, Joyce M. "Designing the 'Perfect' Reference Desk." *Library Journal* (May 15, 1983): 970–72.

Dahlgren, Anders C. "Effective Space Utilization with Whatcha Got." *Iowa Library Quarterly* 22 (Fall 1985): 3–18.

———. *Planning Library Buildings: A Select Bibliography*. Chicago: Library Administration and Management Association, 1990.

———. *Public Library Space Needs: A Planning Outline*. Madison, WI: Department of Public Instruction, 1988.

Depew, John N. *A Library, Media, and Archival Preservation Handbook*. Santa Barbara, CA: ABC-CLIO, 1991.

Dewe, Michael. *Library Buildings: Preparation for Planning Proceedings of the Seminar Held in Aberystwyth, August 10–14, 1987, IFLA Publications*, 48. Munich, Germany: K. G. Saur, 1989.

Dexter Lord, Gayle, Barry Lord, and Lindsay Martin, eds. *Manual of Museum Planning: Sustainable Space, Facilities, and Operations*. Lanham, MD: AltaMira, 2012.

Dodd, Chelsea. "Librarianchels." 2015. http://librarianchels.wordpress.com.

Doucett, Elisabeth. "10 Tips for Tracking Trends; Libraries Can Stay Relevant to Their Users by Strategically Riding the Wave of Societal Trends." *American Libraries* 41, nos. 6 & 7 (2010): 44–47.

Dougherty, Conor. "New Library Technologies Dispense with Librarians." *Wall Street Journal*, October 25, 2010, 98.

Dreyfuss, Joel. "Bionic Books." *Modern Maturity* 43W, no. 5 (2000): 90, 92.

Dubin, Fred. "Mechanical Systems in Libraries." *Library Trends* 36 (Fall 1987): 351–60.

Eberhart, George M. "A Field Guide to Makerspaces." 2015. *AmericanLibrariesmagazine .org.*

Eckelman, Carl A., and Yusuf Z. Erdil. "Test Reports on 15 Models of Bracket-Type Steel Library Bookstacks." *Library Technology Reports* 34, no. 6 (1998): 685–786.

Eisenberg, Anne. "An Express Lane from Camera to Computer." *New York Times*, March 12, 2010.

Farmanfarmaian, Roxane. "Beyond E-Books: Glimpses of the Future." *Publishers Weekly* 248, no. 1 (2001): 56–57.

Fisher, Erin. "Makerspaces Move into Academic Libraries." *ACRL Tech Connect Blog.* Association of College and Research Libraries, 2012.

Fisher, Lawrence M. "Alliance to Send Video-on-Demand on the Web." *New York Times*, August 14, 2000, 3.

Fleischer, Victor, and Jo Ann Calzonetti. "Don't Count on Luck, Be Prepared: Ten Lessons Learned from the 'Great Flood' at the University of Akron's Science and Technology Library." *College & Research Libraries News* 72, no. 2 (2011): 4.

Fleishman, Glenn. "5 New Technologies That Will Change Everything." *PC World*, October 22, 2009.

Fletcher, Janet. *Breaking Down the Barriers—The No Desk Academic Library*. Kensington, New South Wales, Australia: University of New South Wales, 2011.

Fraley, Ruth A., and Carol Lee Anderson. *Library Space Planning: How to Assess, Allocate and Reorganize Collections, Resources, and Physical Facilities*. New York: Neal-Schuman, 1990.

Gaines, Ervin, Marian Huttner, and Frances Peters. "Library Architecture: The Cleveland Experience." *Wilson Library Bulletin* 56, no. 8 (1982): 590–95.

Galluzzi, Anna. "New Public Libraries vis à vis Urban Lifestyles and Global Cities." Milan: International Federation of Library Systems, 2009.

Garbarine, Rachelle. "Residential Real Estate; Computer Program for Apartment Managers." *New York Times*, August 18, 2000, 3.

Garney, Beth A., and Ronald R. Powell. "Electronic Mail Reference Services in the Public Library." *Reference & User Services Quarterly* 39, no. 3 (2000): 249–52.

Genz, Marcella D. "Working the Reference Desk." *Library Trends* 46, no. 3 (Winter 1998): 505–26.

Graff, Susan. "Six Steps to Sustainability." *CRO: Corporate Responsibility Officer*, September 6, 2007.

Grant, Dorothy L., Thomas M. Grant, and Daniel S. Grant. *ADA Compliance Guidelines: California Access Code: Americans with Disabilities Act Title III: California Access Code Title 24*. San Diego: ACR Group, 1994.

Green, Joshua. "No Lectures or Teachers, Just Software." *New York Times*, August 10, 2000, 5.

Green, William R. *The Retail Store: Design and Construction*. New York: Van Nostrand Reinhold, 1991.

"Green Roof." *Scientific American* 298 (2008): 105.

Grimmer, Anne E., and Kay D. Weeks. *The Secretary of the Interior's Standards for the Treatment of Historic Properties: With Guidelines for Preserving, Rehabilitating, Restoring & Reconstructing Historic Buildings*. U.S. Department of the Interior, National Park Services. United States. Washington, DC, 1995.

Guernsey, Lisa. "Bookbag of the Future." *New York Times*, March 2, 2000, 7.

———. "The Library as the Latest Web Venture." *New York Times*, June 15, 2000, 7.

———. "Scan the Headlines? No, Just the Bar Codes." *New York Times*, May 4, 2000, 6.

Habich, Elizabeth Chamberlain. *Moving Library Collections: A Management Handbook*. Westport, CT: Greenwood, 1998.

Hagiu, Andrei, and Bruno Jullien. "Designing a Two-Sided Platform: When to Increase Search Costs?" *Harvard Business Review Working Knowledge for Business Leaders*, 2007. http://idei.fr/sites/default/files/medias/doc/wp/2007/twoside_platform.pdf.

Hagloch, Susan B. *Library Building Projects: Tips for Survival*. Englewood, CO: Libraries Unlimited, 1994.

Hamilton, Anita. "The Public Library Wants to Be Your Office: Shoving Books Aside to Create Community-Centered Coworking Spaces Doesn't Sit Well with Some Library Loyalists." *Fast Company*, August 4, 2014.

Hauke, Petra, and Klaus U. Werner. *The Second Hand Library Building: Sustainable Thinking through Recycling Old Buildings into New Libraries*. San Juan, Puerto Rico: IFLA, 2011.

Hawkins, Brian L., and Patricia Battin, eds. *The Mirage of Continuity: Reconfiguring Academic Information Resources for the 21st Century*. Washington, DC: Council on Library and Information Resources and Association of American Universities, 1998.

Henry, Karen H. *ADA: 10 Steps to Compliance*. Sacramento: California Chamber of Commerce, 1992.

Hlubinka, Michelle, et al. "Makerspace Playbook," School Edition. *Maker Media*, 2013.

Holt, Raymond M. *Planning Library Buildings and Facilities: From Concept to Completion*. Metuchen, NJ: Scarecrow, 1989.

———. "Trends in Public Library Buildings." *Library Trends* 36 (Fall 1987): 267–85.

Holt, Raymond M., and Anders C. Dahlgren. *The Wisconsin Library Building Project Handbook*. Madison, WI: Department of Public Instruction, 1989.

Human Rights Campaign. "Restroom Access for Transgender Employees." 2014. www.hrc.org/resources/entry/restroom-access-for-transgender-employees.

Humphries, A. W. "Designing a Functional Reference Desk: Planning to Completion." *RQ* 33, no. 1 (Fall 1993): 35–41.

IDEO. "Design Thinking for Libraries: A Toolkit for Patron-Centered Design." 2015. http://designthinkingforlibraries.com.

International Union of Architects. "A Guide to Competitive Quality Based Selection of Architects." 2015. www.architecture.com.au/docs/default-source/national-policy/guide-to-quality-based-selection-of-architects.pdf?sfvrsn=2.

Johnson, Larry, Samantha Adams Becker, Victoria Estrada, and Alex Freeman. *NMC Horizon Report: 2015 Higher Education Edition*. Austin, TX: New Media Consortium, 2015.

Jones, Patrick. *Connecting Young Adults and Libraries: A How-to-Do-It Manual*. New York: Neal-Schuman, 1992.

Kavehrad, Mohsen. "Broadband Room Service by Light." *Scientific American* 297, no. 1 (2007): 82–87.

Kim, Bohyun. "Biohackerspace, DIYbio, and Libraries." *ACRL TechConnect Blog*. 2015. http://wp.me/p2zteB-1hE.

Kirkpatrick, David D. "Random House Plans to Establish Exclusively Digital Unit." *New York Times*, July 31, 2000, 5.

Kirwin, William J. "What to Do Until the Architect Comes." *North Carolina Libraries* 39 (Fall 1981): 5–8.

Klasing, Jane P. *Designing and Renovating School Library Media Centers*. Chicago: American Library Association, 1991.

Kolb, Audrey. *A Manual for Small Libraries*. Juneau: Alaska State Library, 1992.

Koontz, Christie. "Retail Interior Layout for Libraries." *Marketing Library Services* 19, no. 1 (2005): 1–4 (electronic articles).

Koontz, Christie, and Dean Jue. "Unlock Your Demographics." *Library Journal* 129, no. 4 (2004): 32–33.

Kroller, Franz. "Standards for Library Building." *Inspel* 16, no. 1 (1982): 40–44.

Langmead, Stephen. *New Library Design: Guidelines to Planning Academic Library Buildings*. Toronto: John Wiley and Sons Canada, 1970.

Larason, Larry, and Judith Schiek Robinson. "The Reference Desk: Service Point or Barrier." *RQ* 23, no. 3 (Spring 1984): 332–38.

Lehmann-Haupt, Christopher. "Creating 'the Last Book' To Hold All the Others." *New York Times*, April 8, 1998, 3.

Leibovich, Lori. "Choosing Quick Hits over the Card Catalog." *New York Times*, August 10, 2000, 6.

Lev-Ram, Michal. "The Race to the Internet of Things." *Fortune* (March 15, 2015): 171.

Lewis, Christopher. "The Americans with Disabilities Act and Its Effect on Public Libraries." *Public Libraries* (January/February 1992): 23–28.

Lewis, Michael. "Boom Box." *The New York Times Magazine*, August 18, 2000.

Lewis, Myron E., and Mark L. Nelson. "How to Work with an Architect." *Wilson Library Bulletin* 57 (September 1982): 44–46.

Library Administration and Management Association. *Library Buildings, Equipment and the ADA: Compliance Issues and Solutions*. Chicago: American Library Association, 1996.

Library of Michigan. *LSCA Builds Michigan Libraries*. Lansing: Library of Michigan, 1986.

Lipow, Anne G. "'In Your Face' Reference Service." *Library Journal* 124, no. 13 (1999): 50–54.

Loder, Michael W. "Libraries with a Future: How Are Academic Library Usage and Green Demands Changing Building Designs?" *College & Research Libraries News* 71, no. 4 (2010): 12.

Lotts, Megan. "Implementing a Culture of Creativity: Pop-up Making Spaces and Participating Events in Academic Libraries." *College & Research Libraries News* 76, no. 2 (2015): 72–75.

Lozano, Camille. "Trans Group Seeks More Bathrooms." San Diego: *The Daily Aztec*. Issue 28 (November 2014): 1.

Lublin, Joann S. "The New Job Is in the Details." *Wall Street Journal* 254 (January 5, 2010): 1.

Lushington, Nolan. "Getting It Right: Evaluating Plans in the Library Building Planning Process." *Library Administration & Management* 7, no. 3 (1993): 159–63.

Lushington, Nolan, and Willis N. Mills. *Libraries Designed for Users: A Planning Handbook*. Syracuse, NY: Gaylord Professional Publications, 1979.

Lynch, Sarah N., and Eugene Mulero. "Dewey? At This Library with a Very Different Outlook, They Don't." *New York Times*, July 14, 2007.

Malenfant, Chuck. "The Information Commons as a Collaborative Workspace." *Reference Services Review* 34, no. 2 (2006): 279–86.

Maltby, Emily. "An ATM for Books Coming Soon: The Most Inclusive Reader's Catalog in the World, at Your Fingertips." *Fortune Small Business, CNN Money.com*, December 14, 2006.

Markoff, John. "I.B.M. Device Raises Storage of Tiny PC's." *New York Times*, June 20, 2000, 3.

Martin, Ron G. *Libraries for the Future: Planning Buildings That Work: Proceedings of the Library Buildings Preconference, June 27–28, 1991, Atlanta, Georgia*. Chicago: American Library Association, 1992.

Massmann, Ann. "The Wood Shelving Dilemma." *Library Resources & Technical Services* 44, no. 4 (2000): 209–14.

May, Francine. *In the Words of the Users: The Role of the Urban Public Library as Place*. Milan: International Federation of Library Systems, 2009.

McCabe, Gerard B. *Operations Handbook for the Small Academic Library*. New York: Greenwood, 1989.

McCarthy, Richard C., and American Institute of Architects. *Designing Better Libraries: Selecting and Working with Building Professionals*. Fort Atkinson, WI: Highsmith, 1995.

Mendels, Pamela. "Universities Adopt Computer Literacy Requirements." *New York Times*, September 29, 1999, 3.

Merrill-Oldham, Jan, and Jutta Reed-Scott. "Library Storage Facilities, Management and Services." In *Systems and Procedures Exchange Center*. Association of Research Libraries, Office of Leadership and Management Services. Flyer 242: 2, 1999.

Mervis, Sybil Stern. "How to Plan a Groundbreaking Ceremony for the Library." *Illinois Libraries* 77, no. 3 (1995): 123–27.

Metcalf, Keyes D. *Planning Academic and Research Library Buildings*. New York: McGraw-Hill, 1965.

———. "Selection of Library Sites." *College & Research Libraries News* 22 (May 1961): 183–92.

Metcalf, Keyes D., Philip D. Leighton, and David C. Weber. *Planning Academic and Research Library Buildings*. Chicago: American Library Association, 1986.

Michaels, Andrea. "Design Today." *Wilson Library Bulletin* 62, no. 8 (1988): 55–57.

Michaels, Andrea, and David Michaels. "Designing for Technology in Today's Libraries." *Computers in Libraries* 12, no. 10 (1992): 8–18.

Miller, Kathy. "3M Unveils a Major New Library Technology System." *Computers in Libraries* 20, no. 1 (January 2000): 26–28.

Moore, Nick. *Measuring the Performance of Public Libraries*. Paris: UNESCO, 1989.

Morgan, Linda. "Patron Preference in Reference Service Points." *RQ* 19, no. 4 (Summer 1980): 373–75.

Mount, Ellis. *Creative Planning of Special Library Facilities*. New York: Haworth, 1988.

Natale, Joe. "Full and Equal Access: Americans with Disabilities Act." *Illinois Libraries* 73, no. 7 (1991): 599–602.

———. "The Next Step: The ADA Self-Evaluation." *Illinois Libraries* 73, no. 7 (1992): 599–602.

National Association to Advance Fat Acceptance (NAAFA). February 2015. www.naafaonline.com/dev2/index.html.

National Institute of Building Sciences (NIBS). *Whole Building Design Guide*. NIBS, 2006.

Nguyen, Daisy. "McDonald's Meets Feng Shui." *San Diego Union Tribune* 17 (March 3, 2008): 1.

Novak, Gloria. "Movable Compact Shelving Systems: Selection and Specifications." *Library Technology Reports* 35, no. 5 (1999): 557–708.

OF Committee. *Summary Report of the Circulation Desk Design Project*, San Jose Public Library. San Jose, CA: San Jose Public Library, 1997: 51.

O'Keefe, Brian. "The Smartest (or the Nuttiest) Futurist on Earth." *Fortune* 155, no. 9 (2007): 60–69.

Oxner, Sheldon R. *How to Select a Contractor*. Omaha, NE: Simmons-Boardman, 1979.

Pack, Todd. "E-Books Take Texts Down a Digital Track." *Orlando Sentinel*, July 15, 2000, 3.

Page, K. *Lighting Program for Libraries*. American Library Association Annual Conference, Chicago, 1995.

Paige, Erinn, Laura Damon-Moore, and Christina Jones. "Library as Incubator Project." 2015. www.libraryasincubatorproject.org/?page_id=9.

Paradis, Richard. "Acoustic Comfort." *National Institute of Building Sciences (NIBS)*. www.wbdg.org/resources/acoustic.php.

Pashia, Angela. "Empty Bowls in the Library: Makerspaces Meet Service." *College & Research Libraries News* 76, no. 2 (2015): 79–82.

Peterson, Kate B., and Janice M. Talcott. "Is Cultural Bias Hurting Your Bottom Line?" *Jewelers Circular Keystone* 169, no. 12 (1998): 86–89.

Pettis, Bre. "Maker Nation: A Celebration of the Resurgence in American Ingenuity." *Popular Mechanics* (2015): 64–97.

Pine, B. Joseph, and James H. Gilmore. *The Experience Economy: Work Is Theatre & Every Business a Stage*. Boston: Harvard Business School Press, 1999.

Piscitello, Shauna. "Feng Shui and Corporate Architecture: It's a Balancing Act." *Corporate Architects eNews*, June 6, 2009. www.slideshare.net/reneekenney/fst -presentation.

Pollet, Dorothy. "New Directions in Library Signage: You Can Get There from Here." *Wilson Library Bulletin* 50 (1976): 456–62.

Popovec, Jennifer. "New Software Makes Green Building Easier." *CONSTRUCTION-MAIL!—the e-mail newsletter from McGraw-Hill Construction*. September 18, 2006.

Prowler, Don. F., FAIA - Don Prowler & Associates. "Whole Building Design Guide." www.wbdg.org/index.php.

"Public Information Kiosk, Inc." 2015. www.pikinc.biz.

Richtel, Matt. "Technology Battle Between Satellite Operators and Cable." www.nytimes .com/2003/04/15/business/media/15BIRD.html.

Rizzo, Joseph P. "Get with the Program! Building a Vision of Place." *Library Journal* 127 (2002): 66–68.

Rohlf, Robert H. "Best-Laid Plans: A Consultant's Constructive Advice." *School Library Journal* 36 (February 1990): 28–31.

———. "New Factor in Planning Public Library Buildings." *Public Libraries* 26 (Summer 1987): 52–53.

———. "The Selection of an Architect." *Public Libraries* 21 (Spring 1982): 5–8.

———. "Setting Your House in Order: Straightforward Advice on Creating User-Friendly Libraries." *American Libraries* (April 1989): 304–6.

Rohlf, Robert H., and David R. Smith. "Public Library Site Selection." *Public Libraries* 24 (Summer 1985): 47–49.

Roose, Tina. "Stress at the Reference Desk." *Library Journal* 114 (September 1, 1989): 166–68.

Rush, Richard D. "The Library of the Future." *The Construction Specifier* (October 1999): 41–44, 58.

Sager, Don, ed. "Changing Perspectives: Joint Use of Facilities by Schools and Public Libraries." *Public Libraries* 38, no. 6 (1999): 355–59.

San Diego City. *Project Management Academy: The Executive Challenge*. City of San Diego (1994): 200.

Sands, Johanna. *Sustainable Library Design*. Sacramento: California State Library, 2002.

Sannwald, William W. *Checklist of Library Building Design Considerations*. 4th ed. Chicago: American Library Association, 2001.

———. *Checklist of Library Building Considerations*. 5th ed. Chicago: American Library Association, 2009.

———. *Event Checklist for Library Groundbreakings and Openings*. San Diego, CA: San Diego Public Library, 1993.

———. "Inspiring Library Spaces from the USA." *PLJ Public Library Journal* 21, no. 5 (2006): 3–7.

———. *Mira Mesa Branch Library Building Program*. San Diego, CA: San Diego Public Library, 1990.

Schibsted, Evantheia. "The Fine Art of Choosing an Architect." *Edutopia: The George Lucas Educational Foundation* no. 3 (2007): 30–32.

Schibsted, Evantheia, and J. J. Sulin. "Going Green: In Chicago's Urban Sprawl, an Environmentally Friendly School Blooms." *Edutopia* 1, no. 9 (2006): 24–28.

Schott, Virginia O. "Site Selection for Rural Public Libraries." *Rural Libraries* 7, no. 2 (1987): 27–59.

Sens, Tom. "Transforming the Academic Library: Considerations When Making Old New Again." *University Business: Models of Efficiency*. 2014. www.universitybusiness.com/transform-academic-library.

Services, M. N. "New Machine Prints a Book in Minutes." *San Diego Union Tribune* 16 (September 26, 2007): A2.

Shelton, John A. *Seismic Safety Standards for Library Shelving, California State Library Manual of Recommended Practice*. Sacramento: California State Library Association, 1990.

Showley, Roger. "Sony's Metreon Is the Disneyland of the New Millennium." *San Diego Union Tribune* (September 5, 1999): 6.

Sidener, Jonathan. "Device Turns the Page on Electronic Displays." *San Diego Union Tribune* 17, no. 2 (2008).

Silberman, Steve. "Ex Libris: The Joys of Curling Up with a Good Digital Reading Device." *Wired* (1998): 98–104.

Silver, Cy H. "Construction Standards for California Public Libraries." *Library Administration & Management* 4, no. 2 (1990): 82–86.

Simon, Matthew J., and George Yourke. "Building a Solid Architect-Client Relationship." *Library Administration & Management* 1 (June 1987): 100–104.

Singh, Rajwant. "Standards and Specifications for Library Buildings." *Lucknow Librarian* 15, no. 2 (1983): 65–73.

Smith, Fran K., and Fred J. Bertolone. *Bringing Interiors to Light*. New York: Watson-Guptill, 1986.

Smith, Lester K. *Planning Library Buildings: From Decision to Design.* Chicago: American Library Association, 1986.

Smith, Lori L. "Is Your Library Plus-Size Friendly?" *American Libraries* 44, nos. 9/10 (2013): 3.

Spencer, Mary Ellen. "Evolving a New Model: The Information Commons." *Reference Services Review* 34, no. 2 (2006): 242–47.

Strauch, Katina. "Selling Points: Shops in the Library." *Wilson Library Bulletin* 68 (February 1994): 45–47.

Strauss, William, and Neil Howe. "The Next 20 Years: How Customer and Workforce Attitudes Will Evolve." *Harvard Business Review* 85, nos. 7–8 (2007): 41–52.

Sullivan, Margaret. "Designing School Libraries for 21st Century Learners." Library Resource Group LLC: 35 Slides, 2009.

Talks about "Makerspaces." Places Where People Can Use 3D Printers and Other Technology. Chattanooga Public Library, 2014. Web page for *Library Journal*: http://lj.libraryjournal.com/2013/05/future-of-libraries/making-room-for-innovation/#_.

Taub, Eric A. "Industry Looks to LED Bulbs for the Home." *New York Times*, May 10, 2009.

Teachout, Terry. "A Hundred Books in Your Pocket: The E-Book Will Transform Reading and Writing." *Wall Street Journal*, January 26, 2006, 14.

Thompson, Clive. "The New Library: Making It without Books." *Wired* 22 (2014).

Thompson, Kalee. "Underground Robot Library." *Popular Science* 280, no. 4 (2012): 2.

Tonner, Shawn. "Designing a Library Service Desk: A Checklist of Considerations." Program Handout. 2000. American Library Association. Richard Fetzer and Elizabeth Titus. Chicago: Library Administration & Management Association, Buildings & Equipment Section, Library Furniture and Equipment Committee: 2.

Tseng, Shu-Hsien. "An Eco-Building, a Healthy Life, and Good Service." *Public Libraries* 46, no. 4 (2007): 50–55.

Turley, L. W., and Ronald E. Milliman. "Atmospheric Effects on Shopping Behavior: A Review of the Experimental Evidence." *Journal of Business Research* 49 (2000): 193–211.

U.S. Department of Energy. "Energy Efficiency and Renewable Energy." www.eere.energy.gov/buildings/info/design/integratedbuilding/passivedaylighting.html.

U.S. Department of Justice. *2010 ADA Standards for Accessible Design.* Washington, DC: Department of Justice, 2010.

U.S. Environmental Protection Agency. "IAQ Tools for Schools Kit—Building Maintenance Checklist." 2006. www.epa.gov/iaq/schools/tfs/building.html.

U.S. Green Building Council. "An Introduction to the U.S. Green Building Council and the LEEDTM Green Building Rating System." 2002. www.usgbc.org.

Veatch, Lamar. "Toward the Environmental Design of Library Buildings." *Library Trends* 36 (February 1987): 45–47.

Waters, Shan. "Types of Store Layouts." In *About.Money.* About.com, 2014.

WBDG Historic Preservation Subcommittee. "Whole Building Design Guide." 2015. www.wbdg.org/design/historic_pres.php.

Weber, Thomas E. "Will Web Publishers Reshape Book Industry or Just Flood Market?" *Wall Street Journal*, July 3, 2000, B1.

Weeks, Kay D. "National Park Service Technical Preservation Services." 2015. www.nps.gov/tps/index.htm.

Wheeler, Joseph L., and Alfred Morton Githens. *The American Public Library: Its Planning and Design with Special Reference to Its Administration and Service.* Chicago: American Library Association, 1941.

Wilgoren, Jodi. "A Revolution in Education Clicks into Place." *New York Times*, March 26, 2000, 4.

Williams, Helene, and Bart Harloe. "The College Library in the 21st Century: Reconfiguring Space for Learning and Engagement." *College & Research Libraries News* 70, no. 9 (2009): 3.

Winters, Willis C., and Bradley A. Waters. "On the Verge of a Revolution: Current Trends in Library Lighting." *Library Trends* 36 (Fall 1987): 327–59.

Wolf, Gary. "Exploring the Unmaterial World." *Wired* 8, no. 6 (2000): 308–19.

Wood, Lamont. "Copper for Fiber: Fiber Optics into the Home, with Help from DSL." *Scientific American* 293, no. 1 (2005): 24.

Yearwood, Simone L. "Catching up with Time: Tips, Tricks, and Best Practices for Library Renovations." *College & Research Libraries News* 76, no. 7 (2015): 362–98.

Yegyazarian, Anush. "Tech.gov: World Wide Library?" *PC World: Technology Perspective Newsletter*, November 10, 2005.